BIRTH OF
A THEOREM

BIRTH OF
A THEOREM

A Mathematical Adventure

CÉDRIC VILLANI

TRANSLATED FROM THE FRENCH

BY MALCOLM DeBEVOISE

ILLUSTRATIONS BY CLAUDE GONDARD

Farrar, Straus and Giroux
New York

Farrar, Straus and Giroux
18 West 18th Street, New York 10011

Grateful acknowledgment is made for permission to reprint illustrations and lyrics from the following previously published material: Photograph of Gribouille copyright © Jean-Pierre Leloir, reproduced by permission of Archives Leloir. Diagrams on page 50 copyright © Springer-Verlag Berlin Heidelberg, used with kind permission of Springer Science+Business Media. "Le Marin et la Rose," lyrics by Jean Huard, music by Claude Pingault, copyright © 1960 Les Éditions Transatlantiques—rights transferred to Première Music Group. All rights reserved. International copyright secured. Used by permission.

Library of Congress Cataloging-in-Publication Data
Villani, Cédric, 1973–
    [Théorème vivant. English]
    Birth of a theorem : a mathematical adventure / Cédric Villani ;
illustrations by Claude Gondard. — First American edition.
        pages   cm
    ISBN 978-0-86547-767-4 (hardback) — ISBN 978-0-374-71023-1 (e-book)
    1. Villani, Cédric, 1973–   2. Mathematicians—France—Biography.
    I. Title.

QA29.V54 A3 2015
510.92—dc23
[B]
                                                        2014031268

Designed by Jonathan D. Lippincott

Preface

I am often asked what it's like to be a mathematician—what a mathematician's daily life is like, how a mathematician's work gets done. In the pages that follow I try to answer these questions.

This book tells the story of a mathematical journey, a quest, from the moment when the decision is made to venture forth into the unknown until the moment when the article announcing a new result—a new *theorem*—is accepted for publication in an international journal.

Far from moving swiftly between these two points, in a straight line, the mathematician moves forward haltingly, along a long and winding road. He meets with obstacles, suffers setbacks, sometimes loses his way. As we all do from time to time.

Apart from a few insignificant details, the story I have told here is in agreement with reality, or at least with reality as I experienced it.

My thanks to Olivier Nora for having encouraged me, on the occasion of a chance encounter, to write this book; thanks to Claire for her careful reading and many helpful suggestions; thanks to Claude for his fine illustrations; thanks to Ariane Fasquelle and the staff at Grasset for grasping at once my purpose in writing this book and for their care in preparing the final manuscript for the typesetter; thanks, finally, to Clément for an unforgettable collaboration, without which this book wouldn't exist.

Cédric Villani

Paris, December 2011

BIRTH OF
A THEOREM

ONE

One o'clock on a Sunday afternoon. Normally the laboratory would be deserted, were it not for two busy mathematicians in need of a quiet place to talk—the office that I've occupied for eight years now on the third floor of a building on the campus of the École Normale Supérieure in Lyon.

I'm seated in a comfortable armchair, insistently tapping my fingers on the large desk in front of me. My fingers are spread apart like the legs of a spider. Just as my piano teacher trained me to do, years ago.

To my left, on a separate table, a computer workstation. To my right a cabinet containing several hundred works of mathematics and physics. Behind me, neatly arranged on long shelves, thousands and thousands of pages of articles, lawfully photocopied back in the days when scientific journals were still printed on paper, and a great many mathematical monographs, unlawfully photocopied back in the days when I didn't make enough money to buy all of the books I wanted. There are also a good three feet of rough drafts of my own work, meticulously archived over many years, and quite as many feet of handwritten notes, the legacy of hours and hours spent listening to research talks. In front of me, Gaspard, my laptop computer, named in honor of Gaspard Monge, the great mathematician and revolutionary. And a stack of pages covered with mathematical symbols—more notes from every one of the eight corners of the world, assembled especially for this occasion.

My partner, Clément Mouhot, stands to one side of the great whiteboard that takes up the entire wall in front of me, marker in hand, eyes sparkling.

"So what's up? Your message was pretty vague."

"My old demon's back again—regularity for the inhomogeneous Boltzmann."

"Conditional regularity? You mean, modulo minimal regularity bounds?"

"No, unconditional."

"Completely? Not even in a perturbative framework? You really think it's possible?"

"Yes, I do. I've been working on it again for a while now and I've made pretty good progress. I have some ideas. But now I'm stuck. I broke the problem down using a series of scale models, but even the simplest one baffles me. I thought I'd gotten a handle on it with a maximum principle argument, but everything fell apart. I need to talk."

"Go on, I'm listening. . . ."

I went on for a long time. About the result I have in mind, the attempts I've made so far, the various pieces I can't fit together, the logical puzzle that so far has defeated me. The Boltzmann equation remains intractable.

Ah, the Boltzmann! The most beautiful equation in the world, as I once described it to a journalist. I fell under its spell when I was young—when I was writing my doctoral thesis. Since then I've studied every aspect of it. It's all there in Boltzmann's equation: statistical physics, time's arrow, fluid mechanics, probability theory, information theory, Fourier analysis, and more. Some people say that I understand the mathematical world of this equation better than anyone alive.

Seven years ago I initiated Clément into this mysterious world when he began his own thesis under my direction. He was eager to learn. Certainly he's the only person who has read everything I've written on Boltzmann's equation. Now Clément is a respected member of the profession, a mathematician in his own right, brilliant, eager to get on with his own research.

Seven years ago I helped him get started; today I'm the one who needs help. The problem I've chosen to work on is exceedingly difficult. I'll never solve it by myself. I've got to be able to explain what I've done so far to someone who knows the theory inside out.

"Let's assume grazing collisions, okay? A model without cutoff. Then the equation behaves like a fractional diffusion, degenerate, of course, but a diffusion just the same, and as soon as you've got bounds on density and temperature you can apply a Moser-style iteration scheme, modified to take nonlocality into account."

"A Moser scheme? Hmmmm . . . Hold on a moment, I need to write this down."

"Yes, a Moser-style scheme. The key is that the Boltzmann operator . . . true, the operator is bilinear, it's not local, but even so it's basically in divergence form—that's what makes the Moser scheme work. You make a nonlinear function change, you raise the power. . . . You need a little more than temperature, of course, there's a matrix of moments of order 2 that have to be controlled. But the positivity is the main thing."

"Sorry, I don't follow—why isn't temperature enough?"

I paused to explain why, at some length. We discussed. We argued. Before long the board was flooded with symbols. Clément was still unsure about the positivity. How can strict positivity be proved without any regularity bound? Is such a thing even imaginable?

"It's not so shocking, when you think about it: collisions produce lower bounds; so does transport, in a confined system. So it makes sense. Unless we're completely missing something, the two effects ought to reinforce each other. Bernt tried a while ago, he gave up. A whole bunch of people have tried, but no one's had any luck so far. Still, it's plausible."

"You're sure that the transport is going to turn out to be positive without regularity? And yet without collisions, you bring over the same density value, it doesn't become more positive—"

"I know, but when you average the velocities, it strengthens the positivity—a little like what happens with the averaging lemmas for kinetic equations. But here we're dealing with positivity, not regularity. No one's really looked at it from this angle before. Which reminds me . . . when was it? That's it! Two years ago, at Princeton, a Chinese postdoc asked me a somewhat similar question. You take a transport equation, in the torus, say. Assuming zero regularity, you want to show that the spatial density becomes strictly positive. Without regularity! He could do it for free transport, and for something more general on small time scales, but for larger times he was stymied. . . . I remember asking other people about it at the time, but no one had a convincing answer."

"Back up. How did he handle the simple free transport case?"

"Free transport" is a piece of jargon that refers to an ideal gas in which the particles do not interact. The model is too simplified to be at all realistic, but you can still learn a lot from it.

"Not sure—but it should be obvious from an explicit solution. Let's try to figure it out, right now. . . ."

Each of us set about reconstructing the argument that this postdoc, Dong Li, must have developed. No big deal, more like a minor exercise in problem solving. But maybe it will help us resolve the great enigma, who knows? And besides, it's a contest—who can come up with the answer first? We scribbled away in silence for a few minutes. I won.

"I think I've got it."

I got up and went over to the board, just like in school when the teacher shows the class how to solve a problem.

"You break down the solution in terms of the replicas of the torus . . . you change the variables in each piece . . . a Jacobian drops out, you use the Lipschitz regularity . . . and finally you end up with convergence in $1/t$. Slow, but it looks about right."

"But then you don't have regularization . . . you get convergence by averaging . . . by averaging. . . ."

Clément was thinking out loud, staring at my calculation. Suddenly his face lit up. In a state of great excitement, he jabbed at the board with his index finger: "But then you'd have to check to see whether that helps with Landau damping!"

I was at a loss for words. Three seconds of silence. A vague feeling this could be important.

Now it was my turn to ask Clément to explain. He didn't know what to say either. He hemmed and hawed, shifting his weight from one foot to the other. Then he said that my solution reminded him of a conversation he'd had three years ago with a Chinese-born mathematician in the United States, Yan Guo, at Brown.

"In Landau damping you want to have relaxation for a reversible equation—"

"Yes, yes, I know. But doesn't interaction play a role? We're not dealing with the Vlasov here, it's just free transport!"

"Okay, maybe you're right, interaction must play a role—in which case . . . the convergence should be exponential. Do you think $1/t$ is optimal?"

"Sounds right to me. What do you think?"

"But what if the regularity was stronger? Wouldn't it be better if it was?"

I groaned. Doubt mixed with concentration, interest with frustration.

We stood in silence, staring at each other, wondering where to go from here. After a while conversation resumed. As fascinating as it is, the weird (and possibly mythical) phenomenon of Landau damping has nothing to do with what we've set out to accomplish. A few more minutes passed and we'd moved on to something else. We talked for a long time. One topic led to another. We took notes, we argued, we got annoyed with each other, we reached agreement about a few things, we prepared a plan of attack. When we left my office a few hours later, Landau damping was nevertheless on our long list of homework assignments.

•

The Boltzmann equation,

$$\frac{\partial f}{\partial t} + v \cdot \nabla_x f = \int_{\mathbb{R}^3} \int_{\mathbb{S}^2} |v - v_*| \left[f(v') f(v'_*) - f(v) f(v_*) \right] dv_* \, d\sigma,$$

discovered around 1870, models the evolution of a rarefied gas made of billions and billions of particles that collide with one another. The statistical distribution of the positions and velocities of these particles is represented by a function f(t, x, v), which at time t indicates the density of particles whose position is (roughly) x and whose velocity is (roughly) v.

Ludwig Boltzmann was the first to express the statistical notion of entropy, or disorder, in a gas:

$$S = -\iint f \, \log f \, dx \, dv.$$

By means of this equation he was able to prove that, moving from an initial arbitrarily fixed state, entropy can only increase over time, never decrease. Left to its own devices, in other words, the gas spontaneously becomes more and more disordered. He also proved that this process is irreversible.

In stating the principle of entropy increase, Boltzmann reformulated a law that had been discovered a few decades earlier, the second law of thermodynamics. *But he did several things that enriched it immeasurably from the conceptual point of view. First, by providing a rigorous proof, he placed an experimentally observed regularity that had been elevated to the status of a natural law on a secure theoretical foundation; next, he introduced an extraordinarily fruitful mathematical interpretation of a mysterious phenomenon; finally, he reconciled microscopic physics—unpredictable, chaotic, and reversible—with macroscopic physics—predictable, stable, and irreversible. These achievements earned Boltzmann a place of honor in the pantheon of theoretical physicists and stimulated an enduring interest in his work among epistemologists and philosophers of science.*

Additionally, Boltzmann defined the equilibrium state of a statistical system as the state of maximum entropy, thus founding a vast field of research known as equilibrium statistical physics. In so doing, he demonstrated that the most disordered state is the most natural state of all.

The triumphant young Boltzmann turned into a tormented old man who took his own life, in 1906. His treatise on the theory of gases appears in retrospect to have been one of the most important scientific works of the nineteenth century. And yet its predictions, though they have been repeatedly confirmed by experiment, still await a satisfactory mathematical explanation. One of the missing pieces of the puzzle is an understanding of the regularity of solutions to the Boltzmann equation. Despite this persistent uncertainty, or perhaps because of it, the Boltzmann equation is now the object of intensive theoretical investigation by an international community of mathematicians, physicists, and engineers who gather by the hundreds at conferences on rarefied gas dynamics and many other meetings every year.

Ludwig Boltzmann

TWO

Landau damping!

In the days following our working session, a confused series of recollections came to me—snatches of conversation, discussions begun but never finished. . . . Plasma physicists have long been used to the idea of Landau damping. But as far as mathematicians are concerned, the phenomenon remains a mystery.

In December 2006 I was visiting Oberwolfach, the legendary institute for mathematical research deep in the heart of the Black Forest, a retreat where mathematicians come and go in an unending ballet of the mind, giving talks on every subject imaginable. No locks on the doors, an open bar, cakes and pastries galore, small wooden cash boxes in which you put payment for food and drinks, tables at which your seat is determined by drawing lots.

One day chance placed me at the same table with two Americans, Robert Glassey and Eric Carlen, both of them authorities on the kinetic theory of gases. The evening before, at the opening of that week's seminar, I had proudly presented a whole batch of new results, and that same morning Eric had given a truly memorable performance, bursting with energy and jam-packed with ideas. The two events, coming one right after the other, were a bit overwhelming for Robert, who confessed to feeling old and worn out. "Time to retire," he sighed. "Retire?" Eric exclaimed in disbelief. There's never been a more exciting time in the theory of gases! "Retire?" I cried.

Just when we are so urgently in need of the wisdom this man has accumulated in his thirty-five years as a professional mathematician!

"Robert, what can you tell me about the mysterious Landau damping effect? Do you think it's real?"

The words "weird" and "strange" stood out in Robert's reply. Yes, Maslov worked on it; yes, there is a paradox of reversibility that seems incompatible with Landau damping; no, it isn't at all clear what's going on. Eric suggested that the effect was chimerical—a product of physicists' fertile imaginations that had no hope of being rigorously formulated in mathematical terms. None of this meant much to me at the time, but I did manage to make a mental note and file it away in a corner of my brain.

Now here we are in 2008, and I don't know anything more about Landau damping than I did two years ago. Clément, on the other hand, had a chance to discuss the matter at length with Yan Guo, one of Robert's younger brothers in mathematics (they both had the same thesis director, twenty years apart). The heart of the difficulty, according to Yan, is that Landau didn't work on Vlasov's original model but on a simplified, *linearized* version. No one knows if what he found also applies to the "true" nonlinear model. Yan is fascinated by this problem—and he's not alone.

Yan Guo

Could Clément and I tackle it? Sure, we could try. But in order to solve a problem, you've got to know at the outset exactly what the problem is! In mathematical research, clearly identifying what it is you are trying to do is a crucial, and often very tricky, first step.

And no matter what our objective might turn out to be, the only thing we'd be sure of to begin with is the Vlasov equation,

$$\frac{\partial f}{\partial t} + v \cdot \nabla_x f - \left(\nabla W * \int f \, dv \right) \cdot \nabla_v f = 0,$$

which determines the statistical properties of plasmas with exquisite precision. Mathematicians, like the poor Lady of Shalott in Tennyson's Arthurian ballad, cannot look at the world directly, only at its reflection—a mathematical reflection. It is therefore in the world of mathematical ideas, governed by logic alone, that we will have to track down Landau. . . .

Neither Clément nor I have ever worked on this equation. But equations belong to everybody. We're going to roll up our sleeves and give it our best shot.

•

Lev Davidovich Landau, a Russian Jew born in 1908, winner of the Nobel Prize in 1962, was one of the greatest theoretical physicists of the twentieth century. Persecuted by the Soviet regime and finally freed from prison through the devoted efforts of his colleagues, he survived to become a towering, almost tyrannical figure in the world of science. With Evgeny Lifshitz he wrote the magisterial ten-volume Course of Theoretical Physics, *still a standard reference today, and made two fundamental contributions to the study of plasma physics in particular: the Landau equation, a sort of little sister to the Boltzmann equation (I studied both in preparing my thesis), and Landau damping, a spontaneous phenomenon of stabilization in plasmas—that is, a return to equilibrium without any increase in entropy, in contrast to the mechanisms described by the* Boltzmann.

Lev Landau

With the physics of gases we are in the realm of Boltzmann: *entropy in-creases, information is lost, the arrow of time points toward the future, the initial state is forgotten; gradually the statistical distribution of neutral particles ap-proaches a state of maximum entropy, the most disordered state possible.*

With plasma physics, on the other hand, we are in the realm of Vlasov: *entropy is constant, information is conserved, there is no arrow of time, the initial state is always remembered; disorder does not increase, and there is no reason for the system to approach one state rather than another.*

Landau had a low opinion of Vlasov, even going so far as to say that al-most all of Vlasov's results were wrong. And yet he adopted Vlasov's model. Landau drew from it a conclusion that Vlasov had completely overlooked, namely, that the electrical forces weakened spontaneously over time without any corresponding increase in entropy or any friction whatsoever. Heresy?

Landau's ingeniously complex mathematical calculation satisfied most physicists, and the so-called damping phenomenon soon came to be named after him. But not everyone was convinced.

THREE

In the hallway, a low table strewn with pages of hastily scribbled notes and a blackboard covered with little drawings. Through the great picture window, a view of a gigantic long-legged black cubist spider, the famous P4 laboratory where experiments are conducted on the most dangerous viruses in the world.

My guest, Freddy Bouchet, gathered up his notes and put them in his bag. We'd spent a good hour talking about his research on the numerical simulation of galaxy formation and the mysterious power of stars to spontaneously organize themselves in stable clusters.

Freddy Bouchet

This phenomenon is not contemplated by Isaac Newton's law of universal gravitation, discovered more than three hundred years ago. And yet when one observes a cluster of stars governed by Newton's law, it does indeed seem that the entire cloud settles into a stable state after a rather long time—an impression that has been confirmed by a great many calculations performed on very powerful computers.

Is it possible, then, to *deduce* this property from the law of universal gravitation? The English astrophysicist Donald Lynden-Bell had no doubt whatsoever about the reality of dynamic stabilization in star clusters. He thought it was a "hard" phenomenon—as hard as, well, an iron meteorite—and gave it the name *violent relaxation*. A splendid oxymoron!

"Violent relaxation, Cédric, is like Landau damping. Except that Landau damping is a perturbative regime and violent relaxation is a highly nonlinear regime."

Freddy was trained in both mathematics and physics, and he has devoted a good part of his professional life to studying such problems. Today he had come to talk to me about one question in particular.

"When you model galaxies, you treat the stars as a fluid—as a gas of stars, in effect. You go from the discrete to the continuous. But how great an error does this approximation entail? Does it depend on the number of stars? In a gas there are a billion billion particles, but in a galaxy there are only a hundred billion stars. How much of a difference does that make?"

Freddy went on in this vein for a long while, raising further questions, explaining recent results, drawing figures and diagrams on the board, noting references. I pointed out the connection between his research and one of my hobbyhorses, the theory of optimal transport inaugurated by Monge. Freddy seemed pleased; it was a profitable conversation for him. For my part, I was thrilled to see Landau damping suddenly make another appearance, scarcely more than a week after my discussion with Clément.

Just as I was coming back to my office after saying goodbye to

Freddy, my neighbor Étienne, who until then had been bustling about, noiselessly filing papers, popped his head into the hallway. With his long gray hair cut in a bob, he looks like an elderly teenager, anticonformist but hardly threatening.

"I didn't really want to say anything, Cédric, but those figures there on the board—I've seen them before."

A plenary speaker at the last International Congress of Mathematicians, member of the French Academy of Sciences, often (and probably rightly) described as the world's best lecturer on mathematics, Étienne Ghys is an institution unto himself. As a staunch advocate of promoting research outside the Paris region, he has spent the past twenty years developing the mathematics laboratory at ENS-Lyon. More than anyone else, he is responsible for turning it into one of the leading centers for geometry in the world. Étienne's charisma is matched only by his grumpiness: he has something to say about everything—and nothing will stop him from saying it.

Étienne Ghys

"You've seen these figures?"

"Yes, that one's from KAM theory. And this one, I know it from somewhere. . . ."

"Where should I look?"

"Well, KAM is found almost everywhere. You start from a completely integrable, quasi-periodic dynamic system and you introduce a small perturbation. There's a problem with small divisors that eliminate certain trajectories, but even so, probabilistically speaking, you've got long-term stability."

"Yes, I know. But what about the figures?"

"Hold on, I'm going to find a good book on the subject for you. But a lot of the figures you see in works on cosmology are usually found in dynamical systems theory."

Very interesting, I'll have to take a look. Maybe it will help me figure out what stabilization is really all about.

That's what I love most of all about our small but very productive laboratory—the way conversation moves from one topic to another, especially when you're talking with someone whose mathematical interests are different from yours. With no disciplinary barriers to get in the way, there are so many new paths to explore!

I didn't have the patience to wait for Étienne to rummage through his vast collection of books, so I rooted around in my own library and came up with a monograph by Alinhac and Gérard on the Nash–Moser theorem. As it happens, I'd made a careful study of this work a few years ago, so I was well aware that the method developed by John Nash and Jürgen Moser is one of the pillars of the Kolmogorov–Arnold–Moser (KAM) theory that Étienne had mentioned. I also knew that Nash–Moser relies on Newton's extraordinary iteration scheme for finding successively better approximations to the roots of real-valued equations—a method that converges unimaginably fast, exponentially exponentially fast!—and that Kolmogorov was able to exploit it with remarkable ingenuity. Frankly, I couldn't see any connection whatever between these things and Landau damping. But who knows, I muttered to myself, perhaps Étienne's intuition will turn out to be correct. . . .

Enough daydreaming! I wedged the book into my backpack

and rushed off to pick up my kids from school, got on the métro and immediately took out a manga from my coat pocket. For a few brief and precious moments life around me disappeared, giving way to a world of supernaturally skilled physicians with surgically recon-structed faces, hardened yakuza who lay down their lives for their children, little girls with huge doe eyes, cruel monsters who suddenly turn into tragic heroes, little boys with blond curls who gradually turn into cruel monsters. . . . A skeptical and tender world, passion-ate, disillusioned, devoid of either prejudice or Manichaean cer-tainties; a world of emotions that strike deep down in the soul and bring tears to the eyes of anyone innocent enough to surrender him-self to them—

Hôtel de Ville! My stop! During the time it took to get here the story had flowed through my brain and through my veins, a small torrent of ink and paper. I felt cleansed through and through.

While I'm reading manga all thoughts of mathematics are sus-pended. It's like hitting a pause button: manga and mathematics don't mix. But what about later, when I'm dreaming at night? What if Landau, after the terrible accident that should have cost him his life, had been operated on by Black Jack? Surely the fiendishly gifted surgeon would have fully restored his powers, and Landau would have resumed his superhuman labors. . . .

For at least a brief time anyway, I was able to forget Étienne's remark and this business about KAM theory. What connection could there possibly be between Kolmogorov and Landau? The moment I got off the métro, the question echoed through my mind over and over again. If there really is a connection, I'll find it.

At the time I had no way of knowing that it would take me more than a year to find the link between the two. Nor could I have sus-pected the fantastic irony that would finally emerge: the figure that caught Étienne's attention, that put him in mind of Kolmogorov, ac-tually illustrates a situation where Landau damping and KAM theory have nothing to do with each other! Étienne's intuition was right, but for the wrong reason—as though Darwin had guessed correctly about the evolution of species by comparing bats and pterodactyls, mistakenly supposing that the two were closely related.

Ten days after the unexpected turn taken by my working session with Clément, a second miraculous coincidence had occurred—and on the same subject! The timing could not have been more fortuitous. Now to take advantage of it.

•

What was the name of that Russian physicist? Just like what happened to me, everyone thought he was dead when they pulled him out from the wreckage. Medically, he *was* dead. An extraordinary story. The Soviet authorities mobilized every resource in order to save an irreplaceable scientist. An appeal for help was even issued to physicians in other countries. The dead man was revived. For weeks the greatest surgeons in the world took turns at his bedside. Four times the man died. Four times life was artificially breathed into him. I've forgotten the details, but I do remember how fascinating it was to read about this struggle against an inadmissible fatality. His tomb was opened up and he was forcibly removed. He resumed his post at the university in Moscow.

[Paul Guimard, *Les choses de la vie*]

•

Newton's law of universal gravitation states that any two bodies are attracted to each other by a force proportional to the product of their masses and inversely proportional to the square of the distance between them:

$$F = \frac{G\,M_1\,M_2}{r^2}.$$

In its classical form, this law does a good job of accounting for the motion of stars in galaxies. But even if Newton's law is simple, the immense number of stars in a galaxy makes it difficult to apply. After all, just because we understand the behavior of individual atoms doesn't mean that we understand the behavior of a human being. . . .

A few years after formulating the law of gravitation, Newton made another extraordinary discovery: an iterative method for calculating the solutions of any equation of the form

$$F(x) = 0.$$

Starting from an approximate solution x_0, you replace the function F by its tangent T_{x_0} at the point $(x_0, F(x_0))$ (more precisely, the equation is linearized around x_0) and solve the approximate equation $T_{x_0}(x) = 0$. This gives a new approximate equation x_1, and you now repeat the same procedure: replace F by its tangent T_{x_1} at x_1, obtain x_2 as the solution of $T_{x_1}(x_2) = 0$, and so on. In exact mathematical notation, the relation that associates x_n with x_{n+1} is

$$x_{n+1} = x_n - [DF(x_n)]^{-1}\,F(x_n).$$

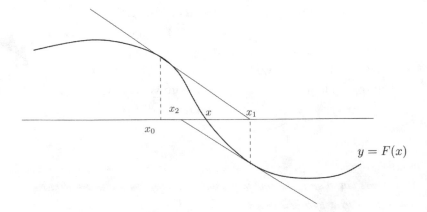

$$y = F(x)$$

The successive approximations x_1, x_2, x_3, . . . obtained in this fashion are incredibly good: they approach the "true" solution with phenomenal swiftness. It is often the case that four or five tries are all that is needed to achieve a precision greater than that of any modern pocket calculator. The Babylonians are said to have used this method four thousand years ago to extract square roots; Newton discovered that it can be used to find not only square roots but the roots of any real-valued equation.

Much later, the same preternaturally rapid convergence made it possible to demonstrate some of the most striking theoretical results of the twentieth century, among them Kolmogorov's stability theorem and Nash's isometric embedding theorem. Single-handedly, Newton's diabolical scheme transcends the artificial distinction between pure and applied mathematics.

Isaac Newton

•

The Russian mathematician Andrei Kolmogorov is a legendary figure in the history of twentieth-century science. In the 1930s, Kolmogorov founded modern probability theory. His theory of turbulence in fluid dynamics, worked out in 1941, remains the starting point for research on this subject today, both for those who seek to corroborate it and for those who seek to disconfirm it. His theory of complexity prefigured the development of artificial intelligence.

Henri Poincaré had convinced his fellow mathematicians that the solar system is intrinsically unstable, and that uncertainty about the position of the planets, however small, makes any prediction of the position of the planets in the distant future impossible. But some seventy years later, in 1954, at the International Congress of Mathematicians in Amsterdam, Kolmogorov presented an astonishing result. Harnessing probabilities and the deterministic equations of mechanics with breathtaking audacity, he argued that the solar system probably *is stable. Instability is possible, as Poincaré correctly saw—but if it occurs, it should occur only rarely.*

Kolmogorov's theorem asserts that if one assumes an exactly soluble mechanical system (in this case, the solar system as Kepler imagined it to be, with the planets endlessly revolving around the sun in regular and unchanging elliptical orbits), and if one then disturbs it ever so slightly (taking into account the gravitational force of attraction, neglected by Kepler), the resulting system remains stable for the great majority of initial conditions.

Kolmogorov's argument was not widely accepted at first. This was mainly because of its complexity, but Kolmogorov's own elliptical style of exposition didn't help matters. Less than a decade later, however, the Russian mathematician Vladimir Arnold and the German mathematician Jürgen Moser, using different approaches, succeeded in providing a complete demonstration, Arnold proving Kolmogorov's original statement of the theorem and Moser a more gen-

Andrei Kolmogorov

eral variant of it. Thus was born KAM theory, which in its turn has given birth to some of the most powerful and surprising results in classical mechanics.

The singular beauty of this theory silenced skeptics, and for the next twenty-five years the solar system was believed to be stable, even if the technical constraints of Kolmogorov's theorem did not correspond exactly to reality. With the work of the French astrophysicist Jacques Laskar in the late 1980s, however, opinion reversed itself once more. But that's another story. . . .

FOUR

The audience holds its breath, the teacher gives the cue, and the young musicians all at once make their bows dance across the strings. . . .

The Suzuki method requires parents to attend their children's group lessons. Here, high in the French Alps, the lessons are given in a grand ski chalet. The main floor is entirely taken up by a stage and rows of seats. What else is there to do but watch—and listen?

We try not to grimace at the most grating noises. Those of us who volunteered yesterday to make fools of ourselves (to our children's great delight) by playing their instruments know full well how difficult it is to make these diabolical contraptions produce the right sound! Today the atmosphere's just right: the adults are in a good mood, the children are happy.

Suzuki method or no Suzuki method, what matters most of all is the teacher, and the one who is helping my son learn to play the cello is really, really talented.

Sitting toward the front, I find myself in almost the same position as my son—devouring Binney and Tremaine's classic work, *Galactic Dynamics*, with the enthusiasm of a small child discovering a new world. I had no idea that the Vlasov equation was so important in astrophysics. Boltzmann's is still the most beautiful equation in the world, but Vlasov's isn't too shabby!

Not only have I come to think better of the Vlasov equation, but all of a sudden the stars in the sky interest me more than they used to. Spiral galaxies and globular clusters always did seem pretty cool. But now that I have a mathematical key to unlock their secrets I find them completely and utterly fascinating.

Since my meeting with Clément three weeks ago I've gone over the calculations again, and now I'm starting to have some ideas of my own, mumbling to myself as I read Binney and Tremaine:

"Don't get it, they say Landau damping is quite different from phase mixing. . . . But aren't they basically the same thing? *Hrmrmrm.*"

A quick glance at the little blond darlings. They're doing fine without me.

"Hmmmm, this calculation's rather interesting. And what's the note at the bottom of the page say? . . . the important thing in the linearized equation is not the spectral analysis, it's the solution of the Cauchy problem. Well, yeah, I guess so—that's common sense! Okay, let's see how they do this . . . hmmmm . . . Fourier transform. Smart move, nothing beats good old Fourier's method. Laplace transform, dispersion relation . . ."

I'm learning quickly, immersing myself in the stuff, absorbing it like a child soaking up a foreign language. Humbly, without pretending to know much of anything, I'm teaching myself basic concepts that physicists have known for half a century. . . .

In the evening I took a break from my studies. Sitting cross-legged in the attic bedroom of the small chalet where we're staying, I devoured the latest collection of short stories by Neil Gaiman, *Fragile Things*— just out in paperback last fall, not yet translated into French. I believe Gaiman is right: we owe it to ourselves to tell stories. Like his tale of a brilliant improvisation on double bass. And of a very old, frail woman recalling her past loves. And of a phoenix that is always rising from its ashes and always being cooked and served for dinner . . .

Closed my eyes, finally, but couldn't fall asleep. Couldn't turn on the light, it's just one room for the whole family. So my brain went haywire. Very old, frail galaxies improvised a story out of Gaiman, the problem kept coming back to life, only to end up being cooked by the mathematicians and served for dinner, over and over again. Stars sprouted inside my head. What exactly was that theorem I wanted to prove?

•

"Crawcrustle," said Jackie Newhouse, aflame, "answer me truly. How long have you been eating the Phoenix?"

"A little over ten thousand years," said Zebediah. "Give or take a few thousand. It's not hard, once you master the trick of it; it's just mastering the trick of it that's hard. But this is the best Phoenix I've ever prepared. Or do I mean, 'This is the best I've ever cooked this Phoenix'?"

"The years!" said Virginia Boote. "They are burning off you!"

"They do that," admitted Zebediah. "You've got to get used to the heat, though, before you eat it. Otherwise you can just burn away."

"Why did I not remember this?" said Augustus Two-Feathers McCoy, through the bright flames that surrounded him. "Why did I not remember that this was how my father went, and his father before him, that each of them went to Heliopolis to eat the Phoenix? And why do I only remember it now?"

[. . .]

"Shall we burn away to nothing?" asked Virginia, now incandescent. "Or shall we burn back to childhood and burn back to ghosts and angels and then come forward again? It does not matter. Oh, Crusty, this is all such fun!"

[Neil Gaiman, "Sunbird"]

•

Fourier analysis is the study of the elementary vibrations of signals. Suppose that we wish to analyze some quantity that varies with time, such as sound, which arises from slight variations in atmospheric pressure. Rather than examine the complex variations of a signal directly, a scientist and politician named Joseph Fourier had the idea in the early nineteenth century of decomposing it into its constituent sine waves—trains of signal pulses known as sinusoids (along with their twins, cosinusoids)—each of which varies in a very simple and repetitive manner.

Joseph Fourier

Each sinusoid is characterized by the amplitude and the frequency of its variations. In a Fourier decomposition, the amplitude measures the relative strength of the corresponding frequency in the signal being analyzed.

Most of the sounds we hear are the result of the superimposition of a multitude of frequencies. A vibration at 440 pulses per second is the musical note A above middle C: the greater its amplitude, the louder it will sound. Double the frequency to 880 pulses and we hear an A one octave higher; triple the frequency to 1320 pulses and the pitch goes to the fifth (which is to say E) in the next higher octave. In the world around us, however, sounds are never pure. They are always made up of a great many frequencies that jointly determine the timbre. When I was preparing for my master's examinations I studied all this in a fascinating course called Music and Mathematics.

Fourier analysis is useful for all sorts of things: decomposing sounds and recording them on a CD, for example, or decomposing images and transmitting them over the Internet, or analyzing variations in the level of the sea and predicting tidal waves. . . .

Victor Hugo delighted in mocking Joseph Fourier, the "little" prefect from the department of Isère whose reputation he felt sure would soon fade— unlike that of the "great" Fourier, the political philosopher Charles Fourier, who he believed would long be remembered by future generations for his utopian socialism. Charles Fourier may not have welcomed the compliment, however. The socialists mistrusted Hugo, and not without reason: he was certainly the greatest writer of his age, but he was no less famous for the changeability of his political opinions; having started out as a monarchist, he became a Bonapartist, an Orléanist, and then a legitimist before exile finally made a republican out of him.

With all due respect to a magnificent author whose works were among my favorites when I was a child, it is indisputable that Joseph Fourier's influence is now much greater than Hugo's ever was. Not only is Fourier's "great mathematical poem" (as Lord Kelvin called it) taught in universities in every country in the world, it is part of the daily lives of billions of people who aren't even aware of it.

•

From a Draft (Dated April 19, 2008)

Formulas will be derived by taking transforms in three variables, x, v, and t. Note that

$$\hat{g}(k) = \int e^{-2i\pi x \cdot k} g(x) \, dx \qquad (k \in \mathbb{Z}^d)$$

$$\tilde{g}(k, \eta) = \int e^{-2i\pi x \cdot k} e^{-2i\pi v \cdot \eta} g(x, v) \, dv \, dx \quad (k \in \mathbb{Z}^d, \eta \in \mathbb{R}^d).$$

Note, too, that

$$(\mathcal{L}g)(\lambda) = \int_0^\infty e^{\lambda t} g(t)\, dt$$

(Laplace transform).

   For the time being it will be assumed that $k \in \mathbb{Z}^d$.

   Taking the Fourier transform in x of the Vlasov equation, we find

$$\frac{\partial \hat{f}}{\partial t} + 2 i \pi (v \cdot k) \hat{f} = 2 i \pi (k \hat{W} \hat{\rho}) \cdot \nabla_v f_0(v).$$

From this, using Duhamel's formula, we deduce

$$\hat{f}(t, k, v) = e^{-2i\pi(v \cdot k)t}\, \hat{f}_i(k, v)$$

$$+ \int_0^t e^{-2i\pi(v \cdot k)(t-\tau)}\, 2 i \pi \hat{W}(k)\, \hat{\rho}(\tau, k)\, k \cdot \nabla_v f_0(v)\, dv.$$

Integrating over v, we find

$$\hat{\rho}(t, k) = \int \hat{f}(t, k, v)\, dv$$

$$= \int e^{-2i\pi(v \cdot k)t} \hat{f}_i(k, v)\, dv + \int_0^t 2 i \pi \hat{W}(k)$$

$$\times \left(\int e^{-2i\pi(v \cdot k)(t-\tau)}\, k \cdot \nabla_v f_0(v)\, dv \right) \hat{\rho}(\tau, k)\, d\tau.$$

(The justification for integrating over v remains to be provided . . . but we can always assume at the outset that the datum has compact support in velocities, then approximate? or truncate. . . .)

   The first term on the right-hand side of the equation on the second line is none other than $\tilde{f}_i(k, kt)$ (this is the same trick used earlier for homogenization by free transport . . .).

   Making lenient assumptions about f_0, we can write, for all $s \in \mathbb{R}$,

$$\int e^{-2i\pi(v \cdot k)s} k \cdot \nabla f_0(v)\, dv = +2 i \pi |k|^2 s \int e^{-2i\pi(v \cdot k)s} f_0(v)\, dv$$

$$= 2 i \pi |k|^2 s\, \tilde{f}_0(ks).$$

Therefore

$$\hat{\rho}(t,k) = \tilde{f}_i(k,kt) - 4\pi^2 \hat{W}(k) \int_0^t |k|^2 (t-\tau) \tilde{f}_0(k(t-\tau)) \hat{\rho}(\tau,k) d\tau.$$

Let us assume

$$p_0(\eta) = 4\pi^2 |\eta| \tilde{f}_0(\eta).$$

(I'm not sure it's a good idea to put 4π here. . . .) In certain cases (such as Maxwellian f_0), p_0 is positive; but generally there is no reason for this to be true. Note that p_0 rapidly decays if $f_0 \in W^{\infty,1}(\mathbb{R}^d)$; it exponentially decays if f_0 is analytic, etc. We obtain, finally,

$$\hat{\rho}(t,k) = \tilde{f}_i(k,kt) - \hat{W}(k) \int_0^t p_0(k(t-\tau)) \hat{\rho}(\tau,k) |k| d\tau.$$

Taking the Laplace transform in $\lambda \in \mathbb{R}$ we obtain, so long as everything is well defined,

$$(\mathcal{L}\hat{\rho})(\lambda,k) = \int_0^\infty e^{\lambda t} \tilde{f}_i(k,kt) dt - \hat{W}(k)$$
$$\times \left(\int_0^\infty e^{\lambda t} p_0(kt) |k| dt \right) (\mathcal{L}\hat{\rho})(\lambda,k);$$

whence we derive

$$(\mathcal{L}\hat{\rho})(\lambda,k) = \frac{\int_0^\infty e^{\lambda t} \tilde{f}_i(k,kt) dt}{1 + \hat{W}(k) Z\left(\dfrac{\lambda}{|k|}\right)},$$

where

$$Z(\lambda) = \int_0^\infty e^{\lambda t} p_0(te) dt, \quad |e| = 1.$$

FIVE

The deafening noise of the cicadas has ceased, but inside the Shugakuin International House the stifling heat lasts late into the night. . . .

Earlier today here at the University of Kyoto I concluded a series of lectures, sponsored by the mathematical research institute and attended by junior professors and graduate students from some fifteen different countries. Today's lecture went well. I began at the appointed hour, or within a minute of it, and finished at the appointed hour, not more than a minute later. Failing to respect a timetable is out of the question in this country. I had to be as punctual as the ferry that took me to Hokkaido last week.

On returning to the visitors' residence this evening I regaled my children with the continuing adventures of Korako, a little Japanese raven. One day, finding himself abandoned by his parents, Korako set off on a long journey through France and Egypt, where he worked in circuses and bazaars, searching for a secret code with his master, a young boy named Arthur. An improvised tale that goes on and on, what my daughter calls an "imaginary story"—her favorite kind, and also the kind that is the most fun for the storyteller.

Then the children went to sleep, and for once I didn't delay in following their good example. After the imaginary story about optimal transport that I told to the audience of budding mathematicians and the imaginary story about the little raven that I made up for my children, I had well earned the right to tell an imaginary story to myself. My brain soon embarked on a fantastic journey of its own.

The tale swept me up and away, and the night went racing by. I woke up with a start, a little after 5:30 in the morning, to that wonderful feeling that lasts only a fraction of a second, when you don't know where you are—not even what continent you're on! I jumped up from the futon and went over to my computer to make a note of the few fragments of the dream I could still hold on to before they completely melted away in the mind's morning fog. The complexity and the confusion of the adventure put me in a good mood: I take such dreams as a sign that my brain is in good working order. They're not as wild or as madly frantic as the ones that David B. records in his comic books, but they are nevertheless convoluted enough that I take great pleasure in trying to remember them.

For several months now I've set Landau damping aside. No real progress yet on a proof, but I have succeeded in clearing a major hurdle: now I know *what* I want to prove. I want to show that *the solution of a nonlinear, spatially periodic, close-to-stable-equilibrium Vlasov equation spontaneously evolves toward another equilibrium*. Even if this way of stating the problem is quite abstract, the abstraction is firmly rooted in reality, in a set of closely related topics of considerable theoretical and practical importance. And even if the problem is simple to state, it's probably difficult to prove. What's more, it asks an original question about a well-known model. So far, so good; I'm very pleased. For the time being I'm keeping Landau damping in the back of my mind. I'll come back to it when classes resume in September.

Beyond the answer to the question (true or false), I very much hope that the proof will tell us many things! Appreciating a theorem in mathematics is rather like watching an episode of *Columbo*: the line of reasoning by which the detective solves the mystery is more important than the identity of the murderer.

In the meantime, there are other passions to indulge: I'm adding an appendix to a paper I wrote two years ago, and I'm making headway on an attempt to combine kinetic equations and Riemannian geometry. Between *local positivity estimates for hypoelliptic equations* and the *kinetic Fokker–Planck equation in Riemannian geometry*, I've got more than enough to keep me busy during these long Japanese nights.

•

OPTIMAL TRANSPORT
AND GEOMETRY

Kyoto, 28 July–1 August 2008

Cédric Villani

ENS-Lyon & Institut Universitaire de France & JSPS

COURSE OUTLINE (5 lectures)
- Basic theory
- The Wasserstein space
- Isoperimetric/Sobolev inequalities
- Concentration of measure
- Stability of a 4th-order curvature condition

Statements will usually be given, but occasionally elements of proof as well.

GROMOV–HAUSDORFF STABILITY OF DUAL
KANTOROVICH PROBLEM
- $(\mathcal{X}_k, d_k) \xrightarrow[k\to\infty]{GH} (\mathcal{X}, d)$ via ε_k-isometries $f_k : \mathcal{X}_k \to \mathcal{X}$
- $c_k(x, y) = d_k(x, y)^2/2$ on $\mathcal{X}_k \times \mathcal{X}_k$
- $\mu_k, \nu_k \in P(\mathcal{X}_k)$ $(f_k)_\# \mu_k \xrightarrow[k\to\infty]{} \mu, (f_k)_\# \nu_k \xrightarrow[k\to\infty]{} \nu$

- $\psi_k : \mathcal{X}_k \to \mathbb{R}$ c_k-convex, $\psi_k^{c_k}(y) = \inf_x [\psi_k(x) + c_k(x, y)]$, achieving sup $\left\{ \int \psi_k^{c_k} d\nu_k - \int \psi_k \, d\mu_k \right\}$

Then, up to extr. $\exists a_k \in \mathbb{R}$ s.t. $(\psi_k - a_k) \circ f_k' \xrightarrow[k \to \infty]{} \psi$, ψ c-convex achieving sup $\left\{ \int \psi^c \, d\nu - \int \psi \, d\mu \right\}$.

Moreover, $\forall x \in \mathcal{X}$, $\limsup\limits_{k \to \infty} f_k \left(\partial_{c_k} \psi_k (f_k'(x)) \right) \subset \partial_c \psi(x)$.

•

The Adventures of Korako (cont'd)

When the moment comes, Korako tosses a stink bomb into the compound, another feat of skill he perfected during his years as a circus performer. Soon the terrible smell makes the guards ill, and Hamad and Tchitchoun set to work filling the air vents with sand.

End of the hunt: the defenses having been breached, the compound is destroyed, Hamad overpowers everyone . . . (long apocalyptic description). Arthur's father has been rescued at last, along with his companion in misfortune, a fellow Egyptologist specializing in hieroglyphics. Their abductors had tried to make them talk about a confidential document—an ancient papyrus containing a secret method for bringing mummies back to life.

The bandits have all been taken prisoner. They are brought to the Madman and told that they are going to be tortured and killed if they don't confess who their leader is. Interrogations follow. Korako is puzzled by the reaction of Arthur's father, who seems to feel at ease; in fact, he seems to know the place—as though he used to live here. Korako secretly listens in on an interrogation and finds his suspicions confirmed: the Madman and Arthur's father already know each other. The next morning he is going to see Arthur and tell him the disturbing news.

[From a summary of the story written up afterward]

•

Notes on a Dream *(August 2, 2008)*

I am an actor in a period film, as well as a member of a ruling family. The historical part of the dream involves both the film and a story in which I am a character, with several simultaneous levels of narration. The prince has absolutely no luck at all. He is constantly being hounded by the crowd, the press. There's a lot of pressure. The king = father of the princess is hatching a plot, something to do with money and a son in hiding. Freedoms are not fully guaranteed. I curse the editorial on the first page of Le Monde. *They have also committed political blunders. But there is grave international concern about the rise in raw materials prices; the Nordic countries, a significant share of whose revenues come from transport, are suffering, particularly Iceland and Greenland. No improvement in sight in any case. I comment on the chances of going to Paris, for example, or at least of meeting famous athletes, they're the real celebrities. I stick holograms containing images of my children on their backs. But a collective suicide has been ordered. Now that the hour has come, I wonder if everyone is present and accounted for. Vincent Beffara isn't here. He played one of the children, but he's no longer suited for the role. The filming has gone on for a long while, and Vincent has grown up in the meantime; instead the same actor is used twice, he doesn't have much to say at the end, there's no problem using a child. I'm very moved, the operation is going to be launched soon. I contemplate the paintings and posters on the walls, which depict the persecution of certain orders of nuns long ago. Nuns belonging to two distinct orders let down their hair before going to their death, despite the general belief that only nuns under the rule of one particular order would die in this condition, that only those nuns had to let down their hair. There is also a painting called* In Praise of Dissidence *or something like that, in which monsters/policemen seize demonstrators who are also vaguely protesters. I give Claire a last kiss. We are very moved. It's almost five o'clock in the morning, the whole family is reunited. I'm going to have to call the agency in charge of transport, disguise my voice, explain that we need explosives and that they can send them here; when they speak of precautions or whatever, I will say (in English): Thank you, I've just been released from a psychiatric hospital (i.e., with these explosives in hand I'll be dangerous). The person at the other end will think that it's a practical joke, he'll go ahead and send everything, and so everything's going to be blown up. Everything is set for 5:30 a.m. I wonder if I'm really going to live my life in an alternative reality, trying to go in another direction, or whether I'll be reborn as a baby, finding myself in limbo for years until my consciousness reemerges . . . I'm rather anxious. . . . Wake up at 5:35—real time!*

SIX

And the days and the nights
　　passed
　　　　in the company of the Problem.

In my sixth-floor walk-up apartment, at the office, in bed asleep . . .

In my armchair, evening after evening, drinking one cup of tea after another after another, exploring paths and subpaths, meticulously noting every possibility, crossing off dead ends from my list as I go along.

One day in October a Korean mathematician, a young woman who had studied under Yan Guo, sent me a manuscript on Landau damping to be considered for publication in a journal of which I am an editor. It was titled "On the existence of exponentially decreasing solutions of the nonlinear Landau damping problem."

For a moment I thought that she and her coauthor had proved the result that I would so dearly like to prove myself, by constructing solutions to the Vlasov equation that spontaneously relax toward an equilibrium! I wrote at once to the editor in chief, saying that I was faced with a conflict of interest and could not in good faith handle the manuscript.

On taking a closer look, however, I realized that they had not come close to doing what I have in mind. They proved only that some damped solutions exist—whereas what needs to be proved is that all solutions are damped! If you know only that some solutions are damped, there's no way of telling whether you're going to come across one of them or not. . . . As it happens, two Italian mathematicians published an article ten years ago proving a fairly similar result, but the authors don't seem to be aware of this earlier work.

No, the Problem hasn't been cracked yet. Besides, it would have been a real disappointment if the solution had turned out to be so simple! An article of thirty pages or so that doesn't resolve any major difficulty is unlikely to do the trick, however good it may be otherwise. Deep down I am convinced that the solution will require completely new tools, which will allow us to look at the problem in a new way.

I need a new norm.

A norm, in mathematical jargon, is a special sort of ruler, or measuring stick, designed for the purpose of estimating the size of some quantity one wishes to investigate. If we want to compare the pluviometry of Brest with that of Bordeaux, for example, should we compare the maximum rainfall for a single day in each place or integrate over the whole year? Comparing maximum quantities involves the L^∞ norm, usually called the supremum (or sup) norm; comparing integrated quantities involves another norm with an equally lovely name, L^1. There are many, many others.

To qualify as a true norm in the mathematical sense, certain conditions must be satisfied. The norm of a sum of two terms, for example, must be less than or equal to the sum of the norms of these terms taken separately. But that still leaves a vast number of norms to choose from.

I need the right norm.

The concept of a norm was formalized more than a century ago. Since then, mathematicians have not stopped inventing new ones. The second-year course I teach at ENS-Lyon is full of norms.

Not only the Lebesque norm but also Sobolev, Hilbert, and Lorentz norms, Besov and Hölder norms, Marcinkiewicz and Lizorkin norms, L^p, $W^{s,p}$, H^s, $L^{p,q}$, $B^{s,p,q}$, $\mathcal{H}^\alpha$, M^p, and $F^{s,p,q}$ norms—and who knows how many more!

But this time none of the norms I'm familiar with seems to be up to the job. I'll just have to come up with a new one myself—pull it out of a great mathematical hat somehow.

The norm of my dreams would be fairly stable under composition with elements close to identity, and capable of accommodating the filamentation typically associated with the Vlasov equation in large time. *Gott im Himmel!* Could such a thing really exist? I tried taking a weighted sup; perhaps I've got to introduce a delay. . . . Clément was saying we need to preserve the memory of elapsed time, in order to permit comparison with the solution of the free transport equation. That's fine with me—but which one is supposed to be taken as the basis for comparison??

While I was rereading the book by Alinhac and Gérard this fall, one exercise in particular caught my eye. *Show that a certain norm W is an algebraic norm.* In other words, show that the norm W of the product of two terms is at most equal to the product of the norms W of these terms taken separately. I've known about this exercise for a long time, but looking at it again I suspected that it might be useful in wrestling with the Problem.

Maybe so—but even if I'm right, we'll still have to modify the evaluation at 0 by inserting a sup, or otherwise an integral. But then that's not going to work very well in the position variable, so we'll have to use another algebraic norm . . . perhaps with Fourier? Or else with . . .

One fruitless attempt after another. Until yesterday. Finally, I think I've found the norm I need. I've been scribbling away for weeks now, evening after evening, page after page, sending the results to Clément as I go along. The machine is cranked up. *Cédurak go!*

•

Let D be the unit disk in $\mathbb{C}$, and $W(D)$ the space of holomorphic functions f on D satisfying

$$\|f\|_{W(D)} = \sum_{n=0}^{\infty} \frac{|f^{(n)}(0)|}{n!} < +\infty.$$

Show that if $f \in W(D)$, and if g is holomorphic near the values taken by f on $\bar{D}$, then $g \circ f \in W(D)$. (Remark that $\| h \|_{W(D)} \leq C \sup_{z \in D}(|h(z)| + |h''(z)|)$, and that $W(D)$ is an algebra; then write

$$f = f_1 + f_2, \text{ with } f_2(z) = \sum_{n>N} \frac{f^{(n)}(0)}{n!} z^n,$$

where N is chosen sufficiently large that the series $\displaystyle\sum_{n=0}^{\infty} \frac{g^{(n)}(f_1)}{n!} f_2^n$ is well defined and converges in $W(D)$.)

[Serge Alinhac and Patrick Gérard,
*Pseudo-Differential Operators and the
Nash–Moser Theorem*
(chapter 3, exercise A.1.a)]

•

Date: Tue, 18 Nov 2008 10:13:41 +0100
From: Clement Mouhot <clement.mouhot@ceremade.dauphine.fr>
To: Cedric Villani <Cedric.VILLANI@umpa.ens-lyon.fr>
Subject: Re: Sunday IHP

I've just seen your last emails, will read them carefully,
I'm getting a lot of flak for trying to use my trick in a
stability theorem for the solution of the transport equation
with small analytic perturbation! More soon! clement

Date: Tue, 18 Nov 2008 16:23:17 +0100
From: Clement Mouhot <clement.mouhot@ceremade.dauphine.fr>
To: Cedric Villani <Cedric.VILLANI@umpa.ens-lyon.fr>
Subject: Re: Sunday IHP

A quick comment after having looked at a paper by Tao
(actually the summary of it that he gives in his blog) on
weak turbulence and the cubic 2d defocusing Schrodinger.

His definition of weak turbulence is: shift of mass to
increasingly higher frequencies asymptotically, and his
definition of strong turbulence is: shift of mass to
increasingly higher frequencies within a finite time. Here's
the conjecture he formulates for his equation: Conjecture.*
(Weak turbulence) There exist smooth solutions u(t,x) to (1)
such that \|u(t)\| _ {H^s({\Bbb T}^2)} goes to infinity as t \to
\infty for any s > 1.

Remains to be seen whether this can also be shown for the
solutions that they're trying to construct (for free
transport, the derivatives in x really blow up). As in our
case they need confinement through the torus apparently in
order to be able to see this phenomenon without dispersion in
the real variable x getting in the way. On the other hand one
thing I don't understand is that he argues the phenomenon is
nonlinear and isn't observed in linear cases. In what we're
looking at it does seem to be found at the linear level . . .

More later, clement

Date: Wed, 19 Nov 2008 00:21:40 +0100
From: Cedric Villani <Cedric.VILLANI@umpa.ens-lyon.fr>

To: Clement Mouhot <clement.mouhot@ceremade.dauphine.fr>
Subject: Re: Sunday IHP

Okay, here's what I've done today. I've added a few comments
to the Estimates file, deleted the first section (which was
obsolete, really) and reorganized various estimates that
were dispersed in various files, so that now pretty much
everything is in a single file.

I don't think we're clear yet about the norm we should be
working in:
- since the equation on \rho in the case of a homogenous
field is integral only in time (!) we have to work in a fixed
norm, which therefore must be _ stable _ under composition
by \Om.
- Fourier seems unavoidable if we're going to be able to
convert the analytic into exponential decay. I don't know
how to do the exponential convergence directly without
Fourier, obviously it must be possible.
- since the change of variable is in (x,v) and the Fourier
transform of \rho is a Dirac in \eta, it looks like what we
need is an analytic norm of the L^2 in k and L^1 in \eta
type.
- but the composition will certainly never be continuous in
an L^1 space, so that can't be right, probably we'll have to
be fairly devious and begin by "integrating" the \etas. That
would leave an L^2 analytic norm in the variable k.

Conclusion: We'll have to go on being devious.

More later,

Cedric

Date: Wed, 19 Nov 2008 00:38:53 +0100

From: Cedric Villani <Cedric.VILLANI@umpa.ens-lyon.fr>

To: Clement Mouhot <clement.mouhot@ceremade.dauphine.fr>

Subject: Re: Sunday IHP

On 19/11/08, 00:21, Cedric Villani wrote:

> Conclusion: We'll have to go on being devious.

Right now I have the impression that in order to find a way
around this difficulty we'll need the theorem on continuity
of composition by Omega for the L^2 analytic norm in Fourier
(without loss of generality . . .), treating \eta as a
parameter. Talk to you tomorrow :-)

Date: Wed, 19 Nov 2008 10:07:14 +0100

From: Cedric Villani <Cedric.VILLANI@umpa.ens-lyon.fr>

To: Clement Mouhot <clement.mouhot@ceremade.dauphine.fr>

Subject: Re: Sunday IHP

After a good night's sleep I see now that it's UNREALISTIC:
composing by Omega will NECESSARILY force us to lose a bit
on lambda (this is already the case when Omega = (1-epsilon)
Id). So we're just going to have to deal with it somehow.
More later. . . .

Cedric

Date: Wed, 19 Nov 2008 13:18:40 +0100

From: Cedric Villani <Cedric.VILLANI@umpa.ens-lyon.fr>

To: Clement Mouhot <clement.mouhot@ceremade.dauphine.fr>

Subject: update

Updated file attached:
I've added subsection 3.2 in which I examine what appears to be a fundamental objection having to do with something we talked about on the phone, the problem of the loss of functional space due to the change of variable. The conclusion is that it isn't lost, but we'll have to be very precise in our estimates of the change in variable.
Cedric

Date: Wed, 19 Nov 2008 14:28:46 +0100
From: Cedric Villani <Cedric.VILLANI@umpa.ens-lyon.fr>
To: Clement Mouhot <clement.mouhot@ceremade.dauphine.fr>
Subject: update

New addition at the end of section 3.2. Things now seem rather promising.

Date: Wed, 19 Nov 2008 18:06:37 +0100
From: Cedric Villani <Cedric.VILLANI@umpa.ens-lyon.fr>
To: Clement Mouhot <clement.mouhot@ceremade.dauphine.fr>
Subject: Re: update

I'm pretty sure that section 5 in its present form is wrong!! The problem arises after the phrase "In assigning powers and factorials": the line that follows seems OK, but in the formula a bit further on the indices don't match up
$(N\_\{k-i+1\}/\{k-i+1\}!$ ought to yield $N\_k/k!$
and not $N\_k/(k+1)!)$

The result seems much too strong. It would mean that in composing by an approximation of the identity the same index

for the analytic norm is preserved. I think we've got to aim
instead at something like

\|f\circ G\| _ \lambda \leq const.

$$\|f\| _ {\lambda \|G\|} \|G\|$$

or something along these lines.

More later,

Cedric

Date: Wed, 19 Nov 2008 22:26:10 +0100

From: Cedric Villani <Cedric.VILLANI@umpa.ens-lyon.fr>

To: Clement Mouhot <clement.mouhot@ceremade.dauphine.fr>

Subject: good news

In the attached version I've gotten rid of the
problematic section 5 (we can always put it back in if
we need to) and in its place I've put calculations on
composition, still using the same analytic variants, which
this time seem to work beautifully as far as composition is
concerned (the formula I had suggested won't do, the right
one turns out to be even simpler, though still of the same
kind).

More later, Cedric

Date: Wed, 19 Nov 2008 23:28:56 +0100

From: Cedric Villani <Cedric.VILLANI@umpa.ens-lyon.fr>

To: Clement Mouhot <clement.mouhot@ceremade.dauphine.fr>

Subject: good news

New version attached. I've checked to see that the usual
calculation can be done with the norm suggested by the

composition rule (section 5.1). It's only slightly more complicated but it seems to give more or less the same result. That's all for today.

Cedric

SEVEN

Headlights suddenly loom out of the darkness at the exit from the parking lot. Momentarily blinded, I approach the driver. It's my third try.

"Excuse me, are you going to Lyon?"

"Uh . . . yes."

"Would you be kind enough to give me a lift? The trains have stopped running!"

The driver hesitates for a split second, glances at her passengers, invites me to take a seat in back. I get in.

"Thanks so much!"

"So you were at the concert?"

"Of course! Wasn't it terrific?"

"Really great, yeah."

"I wouldn't have missed the Têtes Raides' twentieth anniversary tour for anything—but I hate to drive, so I came down by train, thinking I wouldn't have any problem hitching a ride back."

"It's no trouble. I brought along my son and his friend, who's sitting next to you in back."

Hello everybody.

"The pogoing wasn't out of control, there was a lot of room, you weren't getting stepped on all the time. It was pretty relaxed."

"The girls had no reason to complain."

"Oh, some of them love it when it really gets wild!"

Fond memories of one ravishing punk chick in particular, pierced, incredibly full of energy, whom chance threw into my arms one night while dancing at a Pigalle concert in Lyon.

"Nice spider."

"Thank you. I always wear one, it's part of who I am. I have them custom-made in Lyon. Atelier Libellule."

"Are you a musician?"

"No!"

"An artist?"

"A mathematician!"

"A mathematician?"

"Yes—mathematicians *do* exist!"

"What do you work on?"

"Hmmm. Do you *really* want to know?"

"Sure, why not?"

"Yeah, go ahead!"

A deep breath.

"I've developed a synthetic notion of Ricci curvature lower-boundedness in complete, locally compact metric-measure spaces."

"What!?!"

"You must be kidding."

"Not at all. I wrote an article about it that made a pretty big splash in the community."

"What's it about again? Sounds cool."

"Okay, one more time: a synthetic theory of lower-boundedness of Ricci curvature in metric-measure spaces that are separable, complete, and locally compact."

"Wow!"

"What's it good for, anything?"

The ice is broken, we're off. I start talking, explaining, demystifying. Einstein's theory of relativity. Curvature—the cornerstone of non-Euclidean geometry—and the deflection of light rays: if the curvature is positive, the rays get closer the farther they travel; if the curvature is negative, the farther they travel the more they diverge. Curvature is usually expressed using the language of optics, but it can also be expressed using the language of statistical physics: density, entropy, disorder, kinetic energy, minimal energy—that's the discovery I made, with the help of a few other mathematicians. But how can one speak of curvature in a space that's covered with spikes, like the bristles on the head of a hedgehog? Then there's the problem of optimal transport, the subject of my thousand-page book. You encounter it everywhere, economics, engineering, meteorology, computer science, geometry. . . .

I go on and on. The miles fly by.

"We're coming into town now. Where can I drop you off?"

"I live in the first arrondissement—the intellectuals' neighborhood! But anywhere that suits you is fine, really. I'll manage."

"It's no problem at all. Just tell me how to get there and we'll take you home."

"That would be wonderful. Can I give you some money for the toll?"

"No, absolutely not."

"That's very kind of you."

"Before you go, would you write down a mathematical formula for me?"

•

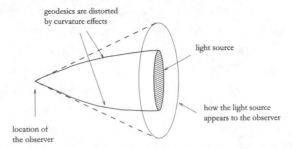

Fig 14.4 – The meaning of distortion coefficients. Because of positive curvature effects, the observer overestimates the surface area of the light source; in a negatively curved world, the observer underestimates it.

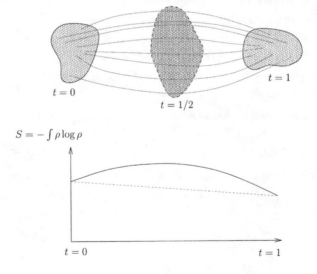

Fig 16.2 – The lazy gas experiment. To go from state 0 to state 1, the lazy gas uses a path of least action. In a nonnegatively curved world, the trajectories of the particles first diverge, then converge, so that at intermediate times the gas can afford to have a lower density (higher entropy).

[From Cédric Villani, *Optimal Transport: Old and New* (Berlin: Springer, 2008), pp. 408, 460, with slight modifications]

EIGHT

Back home with family for the holidays. I've made a lot of progress.

Four computer files, simultaneously updated as we go along, contain everything we have learned about Landau damping. Four files that we have exchanged, added to, corrected, reworked, and sprinkled with notes—marked "NdCM" in boldface for Clément's comments, "NdCV" for mine. Composed in the TEX language developed by the universal master, Donald Knuth, these files are ideally suited for the preliminary maneuvers we're engaged in at the moment.

When we got together again in Lyon a while ago for another working session, Clément complained about an inequality that I had inserted in one of the files:

$$\left\| e^{if} \right\|_\lambda \le e^{\|f\|'_\lambda}.$$

How I could say such a thing was completely beyond him. He had a point, I must admit. I *had* spoken too soon, without thinking it through. At the time this statement seemed to me to be self-evident. Later, after I'd thought about it some more, not only couldn't I say what justification I had for writing the inequality, I no longer understood why it had seemed obvious to me in the first place.

I still don't know why I believed it went without saying. But I've come to realize that it is in fact true! It's true because of Faà di Bruno's formula.

My differential geometry professor at the École Normale Supérieure in Paris long ago introduced me to this formula. It's used for obtaining the successive derivatives of composite functions, and it's unbelievably complicated. I can still picture the scene: lots of murmuring, nervous laughter in the lecture hall; the professor looked at us rather sheepishly, as though he felt he had to apologize, and said, "Don't laugh, it's very useful!"

He was right, Faà di Bruno's formula really *is* very useful: thanks to this formula, my mysterious inequality is true!

But I had to be patient. As Boltzmann, Knuth, and Landau are my witnesses, I swear that for sixteen long years it was of absolutely no use to me whatever. Even the name of the formula's author I had forgotten, unusual though it is.

Somewhere in the back of my mind, however, lurked the thought: *there is a formula for derivatives of composite functions.* Thanks to Google and Wikipedia, it took only a few moments to find the author's name and the formula itself.

The appearance of Faà di Bruno's formula is a sign of the unexpected combinatorial turn that our work has suddenly taken. Ordinarily my drafts are covered with what looks like the sound hole on a cello—the integral sign $\int$ (I've written this symbol so many times that I see it in my mind the moment I start thinking about a problem). But this time they're infested with exponents between parentheses (multiple derivatives: $f^{(4)} = f''''$) and exclamation points (factorials: $16! = 1 \times 2 \times 3 \times \ldots \times 16$).

And so I find myself very much in the holiday spirit even with my work. While the children excitedly open their Christmas presents, I'm hanging exponents on functions like balls on a tree and lining up factorials like upside-down candles!

•

Donald Knuth is the living god of computer science. As one of my colleagues put it, "If Knuth were to walk into the hall in the middle of a lecture, everyone would bow down before him."

Knuth took early retirement from Stanford and turned off his email in order to devote himself full-time to completing his major work, The Art of Computer Programming, *begun almost fifty years earlier. In the meantime, the three volumes already published by 1976 had revolutionized the subject.*

Donald Knuth

In the course of producing these prodigies, Knuth was exasperated by the wretchedly poor graphic quality of the mathematical symbols translated from text file to screen by the software programs that were then commercially available, and resolved to remedy the situation once and for all. Redesigning text editors or fonts wasn't enough: he was determined to rethink the whole business from the ground up. In 1989 he published the first stable version of the TEX typesetting system. The promise of this new universal medium was to be fully realized a decade later, at the beginning of the twenty-first century, when mathematical communication became massively electronic. By the 1990s, however, it was already the standard format used by mathematicians everywhere to compose and share their work.

Knuth's language and its offspring are known as free software, since their source code is available to anyone at no cost. Mathematicians exchange text files, which is to say documents consisting solely of ASCII characters, an alphabet that is recognized by computers throughout the world. Such files contain all the instructions needed to reconstruct natural language text and mathematical formulas down to the smallest detail.

With the invention of this typesetting program, Knuth has probably done more than any other living person to change the daily working lives of mathematicians.

Knuth continually worked to improve the original model. The version numbers he assigned to his program are approximations of π, ever more precisely estimated as the program was gradually perfected: after version 3.14 came version 3.141, then 3.1415, and so on. The current version is 3.1415926; according to the terms of Knuth's will, it will change to π immediately following his death, thus fixing TEX for all eternity.

•

Faà di Bruno's Formula (Arbogast 1800, Faà di Bruno 1855)

$$(f \circ H)^{(n)} = \sum_{\sum_{j=1}^n j m_j = n} \frac{n!}{m_1! \ldots m_n!} \left(f^{(m_1 + \cdots + m_n)} \circ H \right) \prod_{j=1}^n \left(\frac{H^{(j)}}{j!} \right)^{m_j}$$

... which in TEX is written

```
\[(f\circ H)^{(n)} = \sum_{\sum_{j=1}^n j\,m_j = n}
\frac{n!}{m_1!\ldots m_n!}\,
\bigl(f^{m_1 + \ldots + m_n}\circ H\bigr)\,
\prod_{j=1}^n\left(\frac{H^{(j)}}{j!}\right)^{m_j}\]
```

•

Date: Thu, 25 Dec 2008 12:27:14 +0100

From: Cedric Villani <Cedric.VILLANI@umpa.ens-lyon.fr>

To: Clement Mouhot <clement.mouhot@ceremade.dauphine.fr>

Subject: Re: parts 1 and 2, almost done

Here you are, another Christmas present in the form of
part II. It looks very promising, at last everything works
better than one could have hoped on the whole (except that
the exponent loss looks to be at least on the order of a
cubic root of the size of the perturbation, but there's no
reason why it couldn't be recovered using a Newton-style
iterative method). I'm sending you two files: analytic and
scattering, I've stopped fiddling with them for the moment.
It will be necessary to go over them very carefully, but I
think that now our priority should be to make parts 3 and
4 (PDE and interpolation) converge, I suggest that you send
me the PDE part once it looks as though it more or less
holds together even if it still needs some polishing;
that way we can work in parallel on the PDE and
interpolation. (I'll see to putting it into English and
proper form . . .)
And Merry Christmas!
Cedric

Date: Thu, 25 Dec 2008 16:48:04 +0100
From: Clement Mouhot <clement.mouhot@ceremade.dauphine.fr>
To: Cedric Villani <Cedric.VILLANI@umpa.ens-lyon.fr>
Subject: Re: parts 1 and 2, almost done

Merry Christmas and thanks for the presents ;) !!
I'm working on the PDE file to make a complete theorem with
sup and mixed norms, in fact I have high hopes even for the
mixed norm (the scattering is really normed according to
your last file so it seems necessary). As for the
interpolation file, you'll find a draft (in French) of the
revised Nash inequality we need in the version I sent you
earlier, tell me if it needs any more work. More very soon!
Best wishes, Clement

Date: Fri, 26 Dec 2008 17:10:26 +0100
From: Clement Mouhot <clement.mouhot@ceremade.dauphine.fr>
To: Cedric Villani <Cedric.VILLANI@umpa.ens-lyon.fr>
Subject: Re: parts 1 and 2, almost done

Hi,

Here's a preliminary version, in English, of the complete
PDE theorem with your mixed norm, starting on page 15 of the
file. I'm sending it to you now to give you an idea, even if
there are still a few things I have to check regarding the
details of the calculations and the indices . . . , and the
time boundary condition which looks rather odd for the
moment. In any case the mixed norm seems to fit pretty well
with the argument I was making about transfers of normed
derivatives without Fourier. At the beginning of section 4,
in connection with the theorem I mentioned, I've made a few
remarks about why it seems to hold up. On the other hand I'm
still working with a norm with four indices (even it's
definitely a mixed norm by your definition) and for the
moment at least I don't quite see how to get it down to only
three indices . . .
I'll think about it some more.
Best wishes,
Clement

Date: Fri, 26 Dec 2008 20:24:12 +0100
From: Clement Mouhot <clement.mouhot@ceremade.dauphine.fr>
To: Cedric Villani <Cedric.VILLANI@umpa.ens-lyon.fr>
Subject: Re: parts 1 and 2, almost done

What I've been calling the "time boundary condition" is the
fact that since the loss on the index due to scattering is

linear with respect to time (as I put it in the assumption),
that meant there had to be a time boundary if the loss
wasn't to be greater than a certain constant. But now it
seems to me, having looked at your "analytic" file, that the
assumption has to be strengthened, something like a loss
$$
\varepsilon \, \min \{1, (t-s) \}
$$
which allows the loss to remain small for large t and for
s far from t . . .

v. best, clement

NINE

It's pitch-dark; the taxi driver's completely bewildered. His GPS is pointing in a plainly absurd direction: straight ahead into the trees.

I try appealing to his common sense. We've already passed by here once before, obviously the GPS is on the fritz, there's no choice but to explore the surrounding area. In other words, we're lost. The only thing that's certain is that if we follow the machine's instructions, we'll wind up getting stuck in the mud and the melting snow!

In back, the children aren't the least bit worried. My daughter is asleep, worn out by the plane trip and the change in time. My son is watching intently. He's only eight years old but already he's been to Taiwan, Japan, Italy, Australia, and California, so getting lost somewhere in New Jersey isn't about to frighten him. He knows that everything's going to turn out all right.

We drive around some more, see the twinkling lights of civilization in the distance, and then encounter a human being at a bus stop who gives us directions. A GPS has no monopoly on topographic truth.

Finally, the Institute for Advanced Study—the IAS, as everyone calls it—comes into view. A little like a castle rising up in the middle of a forest. We had to go around a large golf course in order to find it. . . .

It is here that Einstein spent the last twenty years of his life. True, by the time he came to America he was no longer the dashing young man who had revolutionized physics in 1905. Nevertheless, his influence on this place was deep and long-lasting, more so even than that of John von Neumann, Kurt Gödel, Hermann Weyl, Robert Oppenheimer, Ernst Kantorowicz, or John Nash—great thinkers all, whose very names send a shiver down the spine.

Their successors include Jean Bourgain, Enrico Bombieri, Freeman Dyson, Edward Witten, Vladimir Voevodsky, and many others. The IAS, more than Harvard, Berkeley, NYU, or any other institution of higher learning, can justly claim to be the earthly temple of mathematics and theoretical physics. Paris, the world capital of mathematics, has many more mathematicians. But at the IAS one finds the distillate, the crème de la crème. Permanent membership in the IAS is perhaps the most prestigious academic post in the world!

And, of course, Princeton University is just next door, with Charles Fefferman and Andrei Okounkov and all the rest. Fields medalists are nothing out of the ordinary at Princeton—you sometimes find yourself seated next to three or four of them at lunch! To say nothing of Andrew Wiles, who never won the Fields Medal but whose popular fame outstripped that of any other mathematician when he broke the spell cast by Fermat's great enigma, which for more than three hundred years had awaited its Prince Charming. If paparazzi specialized in mathematical celebrities they'd camp outside the dining hall at the IAS and come away with a new batch of pictures every day. This is the stuff that dreams are made on. . . .

But first things first: we had to locate our apartment, our home for the next six months, and then get some sleep!

Some people might wonder what there is to do for six months in this very small town. Not me—I've got plenty to do! Above all I need to concentrate. Especially now that I can give my undivided attention to my many mathematical mistresses!

First I've got to wrestle the Landau damping beast to the ground and break its back. I've made good progress so far; the functional framework is firmly established. Two weeks ought to be enough—come on now, time to be done with it! After that I have to finish up the project with Alessio and Ludovic. So far it's eluded us, the damn counterexample we need to prove that for dimension 3 or greater the injectivity domains of an almost spherical Riemannian metric are not necessarily convex. But we're going to find it, and when we do it'll be curtains for the regularity theory of non-Euclidean optimal transport!

If that can be done in two weeks, then I'll have five months left to devote to my great ambition: proving regularity for the Boltzmann! I've brought along all my notes, jotted down in a dozen different countries over the last decade.

Five months might well turn out not to be enough. I was planning to spend two years on the Boltzmann, from last June through the end of my term as a junior member of the Institut Universitaire de France, a five-year appointment during which I have a reduced teaching load in order to give more time to research.

But I keep getting sidetracked. When I began my second book on optimal transport in January 2005, I was determined to limit myself to one hundred fifty pages and to deliver a manuscript sometime in July that same year. In the end it came to a thousand pages, and I didn't finish until June 2008. More than once I thought of stopping midway through and getting back to work on the Boltzmann. But I decided it would be best to persevere. To be honest, I'm not sure I had a choice: it was the book that decided. It couldn't have been any other way.

On stories I really like, I've sometimes fallen behind . . . but it doesn't matter.

As matters stand now, however, I've got only eighteen months left with a reduced teaching load and I still haven't started on what was supposed to be my Big Project. So the invitation to spend a half year in Princeton came at just the right moment. No book to finish, no administrative responsibilities, no courses to teach—I'm going to be able to do mathematics full-time. The only thing I'm required to do is show up for lectures now and then and take part in seminars on geometric analysis, the special theme this year at the IAS School of Mathematics.

Not everyone in the mathematics laboratory at ENS-Lyon was happy about this. They were all counting on me to take over as director of the lab starting in January 2009, exactly the moment I chose to take a leave of absence. Too bad—there are times when one has to put one's own interests first. I've worked for years to help strengthen our group. Once the Princeton interlude is over, I shall be more than happy to work on behalf of the general interest once again.

And then there's the Fields Medal!

The prize whose name no one who covets it dares speak. The highest award there is for mathematicians in their prime, given out every four years on the occasion of the International Congress of Mathematicians to two or three or four mathematicians under the age of forty.

Of course, it's not the only swell prize to be won in mathematics! Indeed, the Abel Prize, the Wolf Prize, and the Kyoto Prize are all probably harder to win than the Fields Medal. But they don't have the same impact or give the same exposure; and since they come at the end of a mathematician's career, they don't serve the same purpose, of recognizing early promise and encouraging continued achievement. The Fields is far more influential.

One tries not to think of it. Thinking of it, trying to win it, would only bring bad luck.

One doesn't even refer to it by name. I'm careful not to mention it in conversation at all. In correspondence I speak simply of the "FM." Whoever I'm writing to knows what I'm talking about.

Last year I won the prize awarded by the European Mathematical Society. In the eyes of many of my colleagues, this was a sign that I was still in the running for the FM. Perhaps the biggest thing in my favor is my range of interests, unusually broad for someone of my generation: analysis, geometry, physics, partial differential equations. And it doesn't hurt that the young Australian prodigy Terry Tao is no longer a candidate, having won the medal at the last ICM in 2006, in Madrid, just after his thirty-first birthday.

But my accomplishments are not entirely immune to criticism.

The conditional convergence theorem for the Boltzmann equation, in which I take such pride, assumes regularity; for the theorem to be perfect, regularity would have to be proved. My work on the theory of Ricci bounds in the weak sense is still in its early stages. The general criterion we've proposed for curvature-dimension is not yet unanimously accepted. And even the great advantage of my versatility carries with it the disadvantage that probably no one mathematician is qualified to judge my achievement as a whole. To have a chance, and also for the sake of my own peace of mind, what I need to do soon—very soon—is to prove a difficult theorem on a significant physical problem.

Then there is the age limit of forty. Right now I'm only thirty-five . . . but with the clarification of the eligibility rule adopted at the last ICM, from now on candidates must be under the age of forty on January 1 of the year of the congress. The moment the new rule was officially announced, I understood what it meant for me: in 2014 I will be too old by three months, so the FM will be mine in 2010— or never. The pressure is enormous!

Since then not a day has gone by without the medal trying to force its way into my mind. Each time it does, I beat it back. Political maneuvering isn't an option, one doesn't openly compete for the Fields Medal; and in any case the identity of the jurors is kept secret. To increase my chances of winning the medal, I mustn't think about it. I must think solely and exclusively about a mathematical problem that will occupy me completely, body and soul. And here at the IAS, I'm in the ideal place to concentrate, following in the footsteps of the giants who came before me.

Just think of it—I'm going to live on Von Neumann Drive!

•

When the stock market crashed in 1929, Louis Bamberger and his sister Caroline Bamberger Fuld could consider themselves lucky. They had amassed a for-

tune from their chain of department stores in Newark, New Jersey, then sold the business six weeks before the stock market collapsed. At a time when the economy lay in ruins, the Bambergers were rich. Very rich.

There is no point being wealthy if one does not put one's wealth to good use. The Bambergers wished to serve a worthy cause, to change society for the better. Their first thought had been to endow a dental school, but soon they were persuaded that their fortune would be best used to establish an institute of theoretical science. Theory was relatively inexpensive. And with all the money at their disposal, why not aim to create the world's foremost institute of theoretical science, an institute whose influence would extend beyond the seas and across the oceans?

In mathematics and theoretical physics, even if researchers don't see eye to eye on everything, they do agree about who the best people are. And once the best people have been identified, well, surely they'll want to pitch in and help make this dream a reality!

After several years of patient negotiation the Bambergers succeeded in luring away the very best, one after another. Einstein came in 1933. Then Gödel. Weyl. Von Neumann. And many more . . . As the political climate in Europe became increasingly unbearable for Jewish scientists and their friends, the world's scientific center of gravity shifted from Germany to the United States. By 1939 the Bambergers' dream had assumed concrete form with the dedication of the Institute's first building, Fuld Hall. On eight hundred acres of land! Adjacent to Princeton University, a prestigious institution almost two hundred years old, itself the beneficiary of another family of philanthropists, the legendarily wealthy Rockefellers. Permanent members of the IAS could look forward to an even more comfortable life than their counterparts at Princeton: no courses to teach, no administrative duties—and extremely generous salaries!

The Institute evolved over the years. Today the School of Natural Sciences is home not only to theoretical physics in all its forms (astrophysics, particle physics, quantum mechanics, string theory, and so on) but also to theoretical biology. Schools of social science and of historical studies came to be added as well, both carrying on the same tradition of excellence.

Mathematicians from around the world come and go, explaining their latest discoveries, hoping to attract the attention of the resident faculty members. Invited visitors, whether they stay for a few months or a few years, must think of only one thing while they are here: producing the best theorems possible—and this under the watchful eye of Albert Einstein himself, whose faintly quizzical smile greets them everywhere they go. In sculptures, photographs, and paintings, Einstein is a constant presence at the Institute.

As a guest of the Institute, your every need has been anticipated. If you are a mathematician, you won't have to worry about anything other than mathematics. If you are accompanied by your family, your children will have been enrolled in school well in advance of your arrival. An army of secretaries stands ready to answer any question and resolve any difficulty. An apartment will be waiting for you only a few minutes from your office. The excellent dining hall will save you the trouble of looking for a restaurant. If you feel like taking a leisurely stroll, you need look no further than the woods that are all around you. Scarcely will you have set foot in the wood-paneled Mathematics–Natural Sciences Library in Fuld Hall than a librarian will introduce herself and help you find the book you are looking for, explaining the card catalogue system that is still in use there, as efficient as it is old-fashioned. The message is unmistakable: Listen, kid, everything you need is right here. Forget about the outside world—your job is to think about mathematics, mathematics, and nothing but mathematics.

If you should ever happen to visit the Institute in the summer, be sure to go see the modernist Historical Studies–Social Science Library overlooking the pond, across from the mathematics building. At night it's deserted. You will feel like an explorer discovering a cave filled with treasures from another age, collections of old maps three feet high and wide, massive dictionaries and encyclopedias heavy enough to be used as doorstops.

Then, on coming out of the library, pause at the edge of the pond: on a late June evening it is the most beautiful place in the world. If you're lucky you'll hear bellowing deer, you'll see the ghostly flickering of fireflies, you'll be mesmerized by the moon's shimmering reflection on the dark water—and you'll sense a spectral aura, the aura of some of the most powerful minds of the twentieth century, forming an invisible mist above the pond.

TEN

Late at night in our new apartment. I'm sitting on the carpeted floor, surrounded by sheets of paper filled with equations and notes. In front of me the big picture window through which the children watch the gray squirrels scurry about outside during the day. Thinking and scribbling away, not saying a word.

In the office, just next door, Claire is watching *Death Note* on her laptop computer. There aren't many cinemas within walking distance of the Institute, so you've got to look for entertainment closer to home. I've praised the film version of this diabolical series to the skies. Now it's Claire's turn to get hooked on it. An opportunity to listen to Japanese again as well.

Today I spoke to Clément on the telephone. We've now gone into high gear. Since I don't have to give any lectures at the IAS, and since as a government-sponsored researcher he doesn't have any teaching obligations either, we can work as hard as we want.

The time difference between France and the United States is a help too. With six hours between us, we can work almost round the clock. If I work until midnight in Princeton, three hours later Clément is in his office in Paris, ready to take over.

Clément has latched onto a particular calculation that involves a pretty neat trick, where you cheat on the existence time of the solution. He has high hopes for it. I don't doubt for a moment that it will be a great help to us going forward.

> In the event, Clément's idea did prove to be of great importance—
> far greater, in fact, than I could have imagined at the time.

But I simply cannot bring myself to believe that by itself it will be enough to save us. We need another estimate.

A new trick.

•

Date: Mon, 12 Jan 2009 17:07:07 -0500
From: Cedric Villani <Cedric.VILLANI@umpa.ens-lyon.fr>
To: Clement Mouhot <clement.mouhot@ceremade.dauphine.fr>
Subject: bad news

So, I haven't had any luck reproducing the transfer of regularity with estimates as good as yours (after conversion in spaces with 3 indices, there's a snag somewhere). I've redone your calculation and found two places where something's not right: (a) the last index on p. 39, l.8 (before "We use here the trivial estimate") seems to me it should be \lambda+2\eta rather than \lambda+\eta; (b) it seems to me impossible that in assumption (5.12) the estimated value does not depend on \kappa (the limits \kappa\to 0 and \kappa\to\infty change the space completely). Conclusion: it seems to me there's a problem. . . .

More later,
Cedric

Date: Mon, 12 Jan 2009 23:19:27 +0100
From: Clement Mouhot <clement.mouhot@ceremade.dauphine.fr>
To: Cedric Villani <Cedric.VILLANI@umpa.ens-lyon.fr>
Subject: Re: bad news

I'll take a closer look tomorrow afternoon. But I agree
about point (a), there surely must be other pbs with indices
as well. As for point (b), what I was thinking of using in
order to say that (5.12) doesn't depend on kappa (for kappa in
a compact set) is the weak dependence with respect to v of
the scattering field $X^{scat}\_{s,t}$: since $\Omega\_{s,t}$
is near to identity within O(t-s), you've got $X^{scat}_{s,t} =
x + O(t-s)$. Whence the fact that any differentiation along v
is "flattened" in the O(t-s)?
Talk to you again soon, clement

Date: Sun, 18 Jan 2009 13:12:44 +0100
From: Clement Mouhot <clement.mouhot@ceremade.dauphine.fr>
To: Cedric Villani <Cedric.VILLANI@umpa.ens-lyon.fr>
Subject: Re: transfer

Hi Cedric,
Since I'm acting as a referee of Jabin's paper on averaging
lemmas (his Porto Ercole course), I checked to see how
closely his calculations agree with ours, and I have the
impression that in the linear estimate the transfer of
regularity has something to do with the averaging lemmas,
but expressed in L^1/L^∞ which seems unusual. For
example if you try to transfer the regularity of x to v
without the gain in x being proportional to (t-s), you're
limited to a gain <1 to have integrability in time, which is
consistent with a limitation of 1/2 in L^2. Another novelty

here in the calculations is that when the gain is
proportional to (t-s) there is no longer a limit 1 . . .
Need to see also if this gain proportional to (t-s) might be
useful in nonlinear regularity theory (your initial
question) . . . Any news on your end?

Best,

Clement

ELEVEN

Every morning I go to the common room in the mathematics building to make myself a cup of tea. Einstein's round, smiling face is
nowhere to be seen here. Watching over the mathematicians in his
stead is André Weil, whose angular features have been memorial-
ized in the form of a bronze bust.

The common room is sparely furnished. Inevitably there's a large
blackboard, in addition to everything you need to make tea and cof-
fee. As well as chessboards and piles of magazines devoted to chess.

One magazine in particular caught my eye, an issue in memory
of Bobby Fischer, the greatest player of all time, who died about a
year ago. Powerless to escape the clutches of paranoia, by the end of
his life he had become an incoherent misanthrope. But beyond the
madness there remain the extraordinary matches of a player whose
abilities have never been equaled, before or since.

In mathematics, as in all other fields of creative endeavor, some
of the greatest minds have suffered a similarly tragic fate.

Paul Erdős, who helped found probabilistic number theory,
was condemned to a life of restless wandering. Amazingly prolific,
Erdős wrote some fifteen hundred articles (a world record), roaming the length and breadth of the globe in his threadbare clothes,
having neither home nor family nor job, only his suitcase, his note-
book, and his genius.

Grigori Perelman, after seven solitary years contemplating the mysteries of Poincaré's famous conjecture, astounded the mathematical world by announcing a solution no one thought possible. Perhaps in order not to spoil the purity of his achievement, Perelman refused the prize of $1 million offered by an American philanthropist—this after having walked away from his post at the Steklov Institute in Saint Petersburg.

Alexander Grothendieck, a living legend, utterly transformed mathematics with the creation of one of the most abstract branches of human thought, and then suddenly resigned from the Institut des Hautes Études Scientifiques, outside Paris. After talks about the possibility of a chair at the Collège de France broke down, he retreated to a small village in the Pyrenees. Once famed for his seductive charm, Grothendieck was fated to pass the rest of his days as a sullen hermit in the grips of madness and a compulsion to write.

Kurt Gödel, the greatest logician of all time, fatally undermined the foundations of mathematics by showing that no axiomatic system rich enough to accommodate arithmetic is complete: any consistent set of axioms contains at least one statement that is neither true (in the sense of being provable) nor false (in the sense of being disprovable). In the last years of his life, ravaged by a severe persecution complex that led him to believe he was in danger of being poisoned, Gödel gradually starved himself to death.

And John Nash, my mathematical hero, revolutionized analysis and geometry with the proof of three theorems in scarcely more than five years before succumbing to paranoid schizophrenia.

There is a fine line, it is often said, between genius and madness. Neither of these concepts is well defined, however. And in the case not only of Grothendieck but also of Gödel and Nash, periods of mental derangement, so far from promoting mathematical productivity, actually precluded it.

Innate versus acquired, a classic debate. Fischer, Grothendieck, Erdős, and Perelman were all Jewish. Of these, Fischer and Erdős were Hungarian. No one who is familiar with the world of science can have failed to notice how many of the most gifted mathematicians and physicists of the twentieth century were Jews, or how many of the greatest geniuses were Hungarian (many of them, but by no means all, Jews). Scientists who worked on the Manhattan Project in the 1940s were fond of saying that Martians really do exist: they have superhuman intelligence, speak an incomprehensible language, and claim to come from a place called Hungary.

Nash, on the other hand, is American through and through, from an old Protestant family. What is more, there was nothing in his ancestry that foretold an exceptional destiny for him. And yet a destiny depends on so many things! The intermingling of genes, the cross-fertilization of ideas, experiences, and chance encounters—all these things have their place in the marvelous, impossibly dramatic lottery of life. Neither genetic inheritance nor environment can explain everything. We should be grateful that this is so.

•

What happens when you gather 200 of the world's most serious scholars, isolate them in a wooded compound, liberate them from all the mundane distractions of university life, and tell them to do their best work? Not much. True, a lot of cutting edge research gets done at the celebrated Institute for Advanced Study near Princeton. Due to the Institute's remarkable hospitality, there is no better place for an academic to sit and think. Yet the problem, according to many fellows, is that the only thing there is to do at the Institute is sit and think. It would be an understatement to call the IAS an Ivory Tower, for there is no more lofty place. Most world-class academic institutions, even the very serious, have a place where a weary bookworm can get a pint and listen to the jukebox. Not so the IAS. Old hands talk about the salad days of the 40s and 50s when the Institute was party central for Princeton's intellectual elite. John von Neumann invented modern computing, but he is also rumored to have cooked up a collection of mind-numbing cocktails that he liberally distributed at wild fetes. Einstein turned physics on its head, but he also took the occasional turn at the fiddle. Taking their clues from the Ancients, the patriarchs of the Institute apparently believed that men (as they would have said) should be well-rounded, engaging in activities high and low, according to the Golden Mean. But now the Apollonian has so overwhelmed the Dionysian at the Institute that, according to many members, even the idea of having a good time is considered only in abstract terms. Walking around the Institute's grounds, you might trip over a Nobel laureate or a Fields medalist. Given the generous support of the Institute, you might even become one. But you can be pretty certain that you won't have adrink or a laugh with either.

[From the liner notes to *Final Report* (1999), a self-produced album by Do Not Erase, the only rock band ever formed at the Institute for Advanced Study]

TWELVE

Saturday evening, dinner together at home.

The whole day was taken up with a trip organized by the Institute for visiting members. A trip to the holiest of shrines for anyone who's enthralled by the story of life: the American Museum of Natural History in New York.

I recall very well my first visit to this museum, almost exactly ten years ago. The excitement of seeing some of the most famous fossils in the world, fossils whose pictures are found in the guides and dictionaries of dinosaurs that I devoured as a teenager, was indescribable.

Today I went back ten years into the past and left my mathematical cares behind for a few hours. Over dinner, however, they caught up with me.

Claire was rather taken aback, seeing my face contorted by tics and twitches.

The proof of Landau damping still hasn't come together. My mind was churning.

What do you have to do, for God's sake, what do you have to do to get a decay through transfer of regularity with respect to position when the velocities have been composed . . . this composition is what introduces a dependence with respect to velocity—but I don't want any velocities!

What a mess.

I scarcely bothered to make conversation, responding in as few words as possible, otherwise by grunts.

"Was it ever cold today! We could have gone sledding. . . . Did you happen to notice the color of the flag at the pond this morning?"

"Hmmm. Red. I think."

Red flag: even if the pond looks frozen, walking on it is prohibited, it's too dangerous. White flag: go ahead, guys, the water's frozen solid, jump and shout, dance on the ice if you like.

And to think that I accepted an invitation to present my results at a statistical physics seminar at Rutgers on January 15! *How* could I have accepted when the proof wasn't complete? What am I going to tell them?

Well, when I got here at the very beginning of the month, I was completely sure I could finish the project in two weeks—max! Fortunately the talk got pushed back by another two weeks! But even with this reprieve, am I going to be ready?? January 29 isn't very far away!! I never thought it would be so hard. No way I could have foreseen the obstacles that lay ahead!

The velocities are the problem, the velocities! When there isn't any dependence with respect to velocity, you can separate the variables by means of a Fourier transform, but when you've got velocities, what can you do? In a nonlinear equation, velocities are obligatory—there's no way I can avoid dealing with them!

"Are you all right? Really, there's no point worrying yourself sick! Relax, take it easy."

"Can't."

"You really seem obsessed."

"Look, I'm on a mission. It's called nonlinear Landau damping."

"I thought you were supposed to be working on the Boltzmann equation. That was your big project, wasn't it? You don't want to lose sight of what you came here to do, do you?"

"Can't be helped. Right now it's Landau damping."

But Landau damping goes on playing the cold, unattainable beauty. I can't get next to her.

. . . *Still, there's that little calculation I did on getting home from the museum—doesn't that give some reason for hope? But man, is it ever complicated! I added two more parameters to the norm. Our norms used to have five regularity indices, which already was the world record—now they're going to have seven! But so what, applying the two indices to a function that doesn't depend on velocity leaves you with the same norm as before, there's no inconsistency. . . . I've got to check the calculation carefully. But if I try to do it right now, it's going to turn out wrong, so let's wait until tomorrow! I'm going to have to do the whole damn thing over again, this time with norms that have got seven indices. Good Lord.*

Seeing how glum I looked, Claire felt sorry for me. Or at least felt she had to do something to cheer me up.

"Look, tomorrow's Sunday. You can spend the day at the office if you like; I'll take care of the little lambkins."

At that moment, nothing in the world could have pleased me more.

•

Date: Sun, 18 Jan 2009 10:28:01 -0500
From: Cedric Villani <Cedric.VILLANI@umpa.ens-lyon.fr>
To: Clement Mouhot <clement.mouhot@ceremade.dauphine.fr>
Subject: Re: transfer

On 01/18/09, 13:12, Clement Mouhot wrote:
> "Any news on your end?"

I'm making progress . . . two steps forward and one step back to begin with, but after a while I became convinced that if we keep going about it your way we won't gain enough in large time. I've found another method that gives us more just on the time variable, it seems to work well enough

except for one thing: it involves somewhat more complicated
spaces, with an additional 2 indices :-) However all the
estimates seem to come out the same for this new family,
still we'll have to check very carefully to be sure. In any
case these are supersubtle tricks, and I think one of the
hearts of the problem. This evening I'll send you a new
version if all goes well, with a few holes yet to be filled
in, and we should be able to begin working in parallel once
more.

Best

Cedric

Date: Sun, 18 Jan 2009 17:28:12 -0500
From: Cedric Villani <Cedric.VILLANI@umpa.ens-lyon.fr>
To: Clement Mouhot <clement.mouhot@ceremade.dauphine.fr>
Subject: Re: transfer

Here's the revised file. For it all to hold together (I
haven't mentioned Newton's scheme yet), we'll have to (i)
check to see that the "bihybrid" norms I've introduced at
the end of section 4 exhibit the same properties as the
"simple" hybrid norms, and that therefore one gets similar
estimates for the characteristics in these norms(!); (ii)
find a way of combining the two distinct effects that are
described in the new section 5; (iii) put all that at the
end of chapter 7 and fill in the missing details to give an
estimate for the total density; (iv) check everything! In
other words, our plates are full. For the moment I suggest
that you go over what I've written and tell me if you see
anything that doesn't look right. I'll let you know later if
I see things that we can clearly work on together at the
same time . . .

A few more points:

I think your estimates for transfer of regularity were flawed, the result was too strong, I wasn't able to reproduce them in the usual norms; on the other hand I did use your transfer strategy in section 5. But when you try to apply it in large time ($t\to\infty$, τ remaining small) the transfer seems to crash, the permissible exponents don't allow the integral to converge over time. I concocted (don't ask me how) a method for getting more integration over time, but this time without getting more regularity. We've got to find a way to combine the two.

More later, best

Cedric

Date: Mon, 19 Jan 2009 00:50:44 -0500

From: Cedric Villani <Cedric.VILLANI@umpa.ens-lyon.fr>

To: Clement Mouhot <clement.mouhot@ceremade.dauphine.fr>

Subject: Re: transfer

I've gone through the file again and cleaned it up a bit, so the attached version should be considered definitive. Going forward I suggest the following division of labor:

- you see to it that Proposition 4.17 and Theorem 6.3 are in good shape, that's asking a lot, I know, but it will have the advantage of forcing you to go over my estimates in sections 4 and 6 again line by line :-) which has to be done, because if we make a mistake calculating the conditions that the exponents must satisfy we're sunk. For the moment in both these sections I've entered some pretty much arbitrary statements as placeholders, with guesstimates that for all I know may turn out to be right but it may also turn out that reality is more complicated. No need to recheck the proofs, but we do have to be absolutely sure about the bounds we end up with, everything else depends on them.

- in the meantime I'll get to work finishing sections 5 and 7 (modulo the input that will come from Theorem 6.3).

- also I'm going to talk to Tremaine tomorrow about the physics part of the intro.

- if you have time to flesh out your comment following, you can incorporate it in the introduction to section 5, where I've already mentioned the connection with averaging lemmas. (Be careful, since we're working in the analytic class it's not wholly obvious that this is an L^1/L^∞ phenomenon??)

If you can get started right away, and assuming all goes well, it ought to be possible to get all this done 2-3 days from now, and only Newton/Nash-Moser will be left to work out in detail. (But I think our priority has to be getting the statements in 4.17 and 6.3 right so we're sure we're not building castles in the air.)

Best
Cedric

Date: Mon, 19 Jan 2009 13:42:27 +0100
From: Clement Mouhot <clement.mouhot@ceremade.dauphine.fr>
To: Cedric Villani <Cedric.VILLANI@umpa.ens-lyon.fr>
Subject: Re: transfer

Hi Cedric,
it's turning into a real monster ;) !!

•

Excerpts from aggregate file-3 (January 18, 2009)

4.7 Bihybrid norms

We shall be led to use the following more complicated norms:

Definition 4.15. We define the space $Z_{(\tau,\tau')}^{(\lambda,\lambda'),\mu;p}$ by

$$\|f\|_{Z_{(\tau,\tau')}^{(\lambda,\lambda'),\mu;p}} = \sum_n \sum_m \frac{1}{n!(n-m)!}$$
$$\times \left\|\left(\lambda(\nabla_v + 2i\pi\tau k)\right)^m \left(\lambda'(\nabla_v + 2i\pi\tau'k)\right)^{n-m}\hat{g}(k,v)\right\|_{L^p(dv)}.$$

$(\dots)$

After trial and error, the best we could do was to recover this decay in the "bihybrid" norms described in subsection 4.7:

Proposition 5.6 (regularity-to-decay estimate in hybrid spaces).
Let $f = f_t(x,v)$, $g = g_t(x,v)$, and

$$\sigma(t,x) = \int_0^t \int f_\tau(x - v(t-\tau),v)\, g_\tau(x - v(t-\tau),v)\, dv\, d\tau.$$

Then

$$\|\sigma(t)\|_{F^{\lambda t + \mu}} \leq \left(\frac{C}{\bar{\lambda} - \lambda}\right) \sup_{0 \leq \tau \leq t} \|f_\tau\|_{Z_\tau^{\bar{\lambda},\mu;1}} \sup_{0 \leq \tau \leq t} \|g_\tau\|_{Z_{(\tau,0)}^{(\lambda,\bar{\lambda}-\lambda),\mu}}.$$

THIRTEEN

Princeton
January 21, 2009

Thanks to the rabbit I pulled out of my hat on my returning from the museum the other evening, I've been able to get back on track. But today I'm filled with a strange mixture of optimism and dread. Got around one major roadblock: made a few explicit calculations and eventually figured out how to manage a term that had gotten too big—that much gives me hope. At the same time, the complexity of the mathematical landscape that's now opened up makes my head spin if I think about it for more than a few moments.

Could it really be that Vlasov's splendid equation, which I thought I was beginning to get a handle on, operates only by fits and starts? On paper, at least, it looks as though sometimes the response to external perturbations suddenly occurs very, very quickly. I've never heard of such a thing; it's not in any of the articles and books that I've read. But in any case we're making progress.

•

Date: Wed, 21 Jan 2009 23:44:49 -0500
From: Cedric Villani <Cedric.VILLANI@umpa.ens-lyon.fr>
To: Clement Mouhot <clement.mouhot@ceremade.dauphine.fr>
Subject: !!

There we go, finally, after hours of floundering miserably I'm pretty sure I've figured out why the O(t) I was complaining about on the phone today gets canceled. It's a MONSTER!

Apparently the problem isn't in the bilinear estimates or in Moser's scheme, it occurs at the level of the "Gronwall" equation in which \rho is estimated as a function of itself . . . the point is that we're dealing with something like

u(t) \leq source + \int _ 0^t a(s,t) u(s) ds

where u(t) is a bound on \|\rho(t)\|. If \int _ 0^t a(s,t) ds = O(1), everything's fine. The problem is that \int _ 0^t a(s,t) can also be equal to O(t) (really not an obstacle in and of itself, I've chosen the most perfect possible cases and this can always happen). But when it does happen, it's at a point strictly internal to [0,t], in the middle somewhere (as in the case where you've got k and \ell such that 0 = (k+\ell)/2); or around 2/3 if you have 0− (2/3)k + \ell/3, etc. But then the recurrence equation on u(s) becomes

u(t) \leq source + epsilon t u(t/2)

and the solutions to this thing turn out not to be bounded, they slowly decay! (subexponential) But since the norm on \rho restricts exponential decay, you do in fact end up getting this decay. . . . .

Whipping this thing into shape looks pretty daunting (resonances will all have to be catalogued, basically). That's my job for tomorrow. In any case none of this

relieves us of the obligation to verify the properties of the bihybrid norms.
Best
Cedric

Date: Wed, 21 Jan 2009 09:25:21 +0100
From: Clement Mouhot <clement.mouhot@ceremade.dauphine.fr>
To: Cedric Villani <Cedric.VILLANI@umpa.ens-lyon.fr>
Subject: Re: !!

It really does look like we've got a monster on our hands! As for me, I've looked at the Nash-Moser part and I agree it seems unlikely that the factor t can be imported into it . . . On the other hand if I correctly understand the argument concerning the bound on u(t) it is absolutely necessary that the point s where a(s,t) is large remain uniformly at a strictly positive distance from t . . . Another thing is that we would therefore be dealing with a subexponential time bound in solving the nonlinear problem. And to have it eaten up by the norm on \rho we'd have to accept losing a bit on its index, which in my opinion must absolutely be avoided in the Nash-Moser part . . . ?
best, clement

FOURTEEN

Darkness! I've got to be in the dark. I have to be alone in the dark. Where? The children's room, shutters closed!

Regularization. Newton's scheme. Exponential constants. Everything was swirling around and around inside my head.

I'd brought the children back home from school and immediately locked myself away in their room. Had to kickstart my brain and put it into high gear. Tomorrow's my talk at Rutgers, and the proof still hasn't come together.

I've got to be by myself. I've got to walk around and think. Think hard. It's urgent!

Claire has put up with worse than this without so much as a word. But here I was walking around in circles, alone in a dark room, while she was in the kitchen making dinner. It was a bit much.

"This is getting really weird!!"

I didn't respond. Circuits overloaded, too many mathematical signals trying to get through. And the pressure of having to make something happen—fast. Even so, when dinner was served I came out to eat with the family, then went back to work for the entire evening. One calculation in particular that I thought for sure was correct turned out to be trouble; somehow I got something wrong. How serious an error was it? I had to find out.

Finally quit around two o'clock in the morning, now to bed. I have a feeling everything's going to be fine after all.

•

Date: Thu, 29 Jan 2009 02:00:55 -0500
From: Cedric Villani <Cedric.VILLANI@umpa.ens-lyon.fr>
To: Clement Mouhot <clement.mouhot@ceremade.dauphine.fr>
Subject: aggregate-10

!!!! I think we've now got the missing pieces.

-First, I finally figured out (unless there's a mistake) what needs to be done to lose an epsilon as small as one likes (even if it means losing a very large constant, either exponential or exponential squared in 1/epsilon). This drops out of a perfectly diabolical calculation that for right now I've simply sketched at the end of section 6. It seems totally miraculous but everything falls in place just the way it should, looks like it must be right.

- Next, I think I've also figured out exactly where we lose on the characteristics and the scattering. We're going to have to redo all the calculations in this section, which won't be any fun . . . I've inserted a few comments in a subsection at the end of this section.

With that, I think we've now got everything we need to feed the Nash-Moser. Tomorrow (Thursday) I'm not here. Here's what I suggest we do after that: I go back over section 6 and subexponential growth again while you tackle the scattering estimates, which aren't too depressing. Let's aim to have everything except the last section revised and ready to go early next week sometime. Will that work for you?

Best

Cedric

FIFTEEN

The dreaded day had finally come. The day of my talk to the mathematical physics seminar at Rutgers University, not quite twenty miles up the road from Princeton. I got a ride this morning with Eric Carlen and Joel Lebowitz, both of whom live in Princeton and work at Rutgers.

This was my second visit there. The first one was a couple of years ago for a gathering in memory of the late Martin Kruskal, the inventor of solitons, a great mind. I still vividly recall the amusing anecdotes recounted by the speakers that day. One in particular of Kruskal and two colleagues talking in an elevator, so deeply immersed in conversation that they went up and down for twenty minutes, oblivious to the people getting on and off.

Today was less entertaining. I was under a lot of pressure!

Usually a seminar talk is devoted to presenting a result that has been meticulously checked to be sure that everything is correct. That's what I've always done in the past. Today was different: not only hadn't the proof I was about to present been scrutinized down to the least detail, it wasn't even complete.

Last night, of course, I had convinced myself that everything was in good shape, and that it remained only to write up the last part. But this morning my doubts returned. Before being dispelled once more. Then, in the car, I had to fight them off again.

When the time came to give my talk, I really was convinced that everything was fine. Self-delusion? I didn't go into a lot of mathematical detail. Instead I spoke at some length about the significance of the problem and its mathematical interpretation, and unveiled my vaunted norm, the complexity of which made my listeners gasp—even though I had limited myself to presenting the version with five indices, not the one with seven. . . .

Afterward about a dozen of us sat down together for lunch. The conversation flowed easily. In the audience earlier one could not have failed to notice Michael Kiessling, a giant elf of a man, eyes sparkling with an air of impish exuberance. Now, at the table, Michael was telling me with his usual infectious enthusiasm about how as a young man he fell in love with plasma physics, screening, the plasma wave echo, quasi-linear theory, and so on.

Michael Kiessling

His mention of the plasma echo immediately concentrated my full attention. What a lovely experiment! You begin by preparing a plasma, which is to say a gas in which the electrons have been separated from the nuclei, so that the plasma is at rest. Then you disturb this motionless state by briefly applying an electric field, which is to say a "pulse" (actually a train of pulses) that excites the gas. Once the current that has been propagated in this manner has faded away, a second field is applied. You wait for the current to fade away a second time. It is at this point that the miracle occurs: if the two pulses have been well modulated, you will observe a spontaneous response, at a precise instant. This response is called the *echo*. . . .

Spooky, huh? One utters an (electrical) cry in the plasma wilderness, then a second cry (in a different pitch), and a few moments later the plasma responds with a cry of its own (in yet another pitch)!

All this brought to mind some calculations I did a few days ago: a temporal resonance . . . the plasma reacting at certain quite specific moments . . . I thought that I'd lost my mind—but perhaps it's the same thing as the echo phenomenon that plasma physicists have known about for years?

I made a mental note to think about this some more. For the moment, however, I was content to make small talk with my hosts. Who have you got in your department right now? Recruited any good people lately? Why yes, things are going very well, there's So-and So and So-and-So, also So-and-So, and So-and-So—

This last name gave me a start.

"What!? Vladimir Scheffer works here!!"

"Yes—though it's been ages since anyone's seen him. Why do you ask, Cédric? Do you know his work?"

"Yes, of course! I gave a Bourbaki seminar just last year on his famous existence theorem for paradoxical solutions of Euler's equations. . . . I've got to meet him!"

"He's not around much, no one's talked to him in a long time. I can try to track him down after lunch if you'd like."

In the event, Joel did manage to get in touch with Scheffer. He agreed to join us later that afternoon.

I won't forget my encounter with him anytime soon.

Scheffer apologized at length for not having been able to come earlier, something to do with administrative duties that seemed to involve forestalling threats of legal action against the university by disgruntled students. It wasn't very clear.

After a while Scheffer and I excused ourselves to talk privately in a small room with a blackboard next door to Joel's office.

"I gave a Bourbaki seminar on your work. Here, I've printed out the text for you! It's in French, but perhaps you'll be able to get something out of it. I explain in great detail how your existence theorem for paradoxical solutions was improved and simplified by De Lellis and Székelyhidi."

"Ah, this is very interesting, thank you."

"I wanted to ask you, how did you ever come up with the idea of constructing these incredible solutions?"

"Well, it's really very simple. In my thesis I had shown that there exist impossible objects, things that shouldn't exist in our world. Here's the method."

He drew a few humps on the board, and a sort of four-pointed star. I recognized the figure.

"Yes, of course, that's Tartar's T_4 configuration!"

Luc Tartar

"Really? Well, maybe, I don't know. In any case I did that in order to construct impossible solutions to certain elliptical equations. And then I realized that there was a general formula."

He explained the formula.

"Yes, of course, that's Gromov's method of convex integration!"

"Really? No, I don't think so. What I was doing is much simpler. The construction is very straightforward, it works because you're in the convex envelope and you can express the approximate solution each time as a convex combination, and then. . . ."

But these are the main elements of convex integration theory! Did this guy really rediscover everything all by himself, without any idea what other people had already done? Where was he living? On Mars?

"So what about fluid mechanics?"

"Oh, right! I'd heard Mandelbrot give a talk, and I said to myself, I'd like to do something similar. So I began to study Euler's equations from the fractal point of view, and I realized that I could reinterpret the same kind of things I'd done in my thesis. But it was complicated."

I was on the edge of my seat, listening intently. Then, after making two or three general remarks, he abruptly stopped.

"I'm sorry, I've got to go now. I use public transportation, just now with the snow it's very slippery and my balance isn't very good, it's a rather long walk home from the bus stop, and . . ."

Once he'd finally gone through all the reasons why he had to leave right away, the monologue came to an end. The time we'd actually spent discussing mathematics was no more than five minutes, from which I learned nothing. To think that this was the same man who is responsible for the most amazing theorem in all of fluid mechanics! Living proof that having a superior mind is no guarantee of being able to communicate.

I returned to Joel's office and told him about my meeting with Scheffer. I said that I was sorry it had been so brief.

"Cédric, five minutes with Vlad is about as much time as anyone in the department has spent talking with him in the last five years."

An experience that will remain forever engraved in my memory. . . . But now it was time to go back to thinking about Landau damping.

On the way home, my doubts returned once more.

Finally I realized the proof isn't going to work.

The talk at Rutgers marked a turning point in my quest. To announce results you have not yet proved is a grave sin, a violation of the bond of trust that unites a speaker and his audience. My back was to the wall: if I was ever to make up for my transgression, I had to prove the result I had just announced.

John Nash, my mathematical hero, is said to have regularly put himself under fantastic pressure by announcing results that he did not yet know how to prove. There is no question he did this at least once, in the case of the isometric embedding theorem.

On my way home that day I began to feel something like the pressure Nash must have felt. The terrible sense of urgency that now hung over me was not going to go away until I succeeded, one way or another, in completing this proof. I *had* to complete it—or else forever be disgraced!!

•

Imagine you're walking through the woods on a peaceful summer's afternoon. You pause at the edge of a pond. Everything is perfectly calm, not the slightest breeze.

Suddenly the surface of the pond becomes agitated, as though seized by convulsions; a few moments later, it is sucked down into a roaring whirlpool.

And then, a few moments after that, everything is calm once more. Still not a breath of air, not even a ripple on the surface from a fish swimming beneath it. So what happened?

The Scheffer–Shnirelman paradox, surely the most astonishing result in all of fluid mechanics, proves that such a monstrosity is possible, at least in the mathematical world.

It is not based on an exotic model of quantum probabilities or dark energy or anything of that sort. It rests on the incompressible Euler equations, the oldest of all partial differential equations, used by mathematicians and physicists everywhere to describe a perfectly incompressible fluid without any internal friction.

It has been more than two hundred fifty years since Euler derived his fundamental equations, and yet not all of their mysteries have been penetrated. Indeed, they are still considered to mark out one of the most treacherous regions of the mathematical world. When the Clay Mathematics Institute set seven "millennium problems" in 2000, offering $1 million apiece for their solution, it did not hesitate to include the regularity of solutions to the Navier–Stokes equations. It was very careful, however, to avoid any mention of Euler's equations—a far greater and more terrifying beast.

And yet at first glance Euler's equations seem so simple, so innocent, utterly devoid of guile or cunning. No need to model variations in density or to grapple with the enigmas of viscosity. One has only to write down the classical laws of conservation: conservation of mass, quantity of motion, and energy.

But then . . . suddenly, in 1993, Scheffer showed that Euler's equations in the plane are consistent with the spontaneous creation of energy! Energy created from nothing! No one has ever seen such bizarre behavior in fluids in the natural world! All the more reason, then, to suspect that Euler's equations hold still more surprises in store for us. Big surprises.

Scheffer's proof is a stunning feat of mathematical virtuosity, as obscure as it is difficult. I doubt that anyone other than its author has read it carefully from beginning to end, and I am certain that no one could reconstruct its reasoning, unaided, in every detail.

There was more to come. Four years later, in 1997, the Russian-born mathematician Alexander Shnirelman, renowned for his originality, presented a new proof of this staggering result. Shortly afterward Shnirelman proposed a physically realistic criterion for solutions to Euler's equations that would prohibit pathological phenomena of the sort Scheffer had discovered.

Alas! A few years ago, two brilliant young mathematicians, Camillo De

Lellis, an Italian, and László Székelyhidi, a Hungarian, proved a general and still more shocking theorem that showed, among other things, that Shnirelman's criterion was powerless to resolve the paradox. Additionally, using the techniques of convex integration, they were able to develop a new method for producing these "wild" solutions, an elegant procedure that grew out of earlier research by a number of mathematicians, including Vladimir Šverák, Stefan Müller, and Bernd Kirchheim. Thanks to De Lellis and Székelyhidi, we now realize that even less is known about Euler's equations than we thought.

And what we thought we knew wasn't much to begin with.

•

Excerpt from the 2008 Bourbaki Seminar Talk

Theorem [Scheffer (1993), Shnirelman (1997)]. There exists a nonzero weak solution of the incompressible Euler equations in dimension 2,

$$\frac{\partial v}{\partial t} + \nabla \cdot (v \otimes v) + \nabla p = f, \quad \nabla \cdot v = 0,$$

without forcing ($f \equiv 0$), with compact support in space-time.

Theorem [De Lellis and Székelyhidi (2007, 2008)]. Let Ω be an open interval of $\mathbb{R}^n$, $T > 0$, and $\overline{e}$ a uniformly continuous function $\Omega \times]0, T[\to]0, +\infty[$, with $\overline{e} \in L^\infty(]0, T[; L^1(\Omega))$. Then for all $\eta > 0$ there exists a weak solution (v, p) of the Euler equations, without forcing ($f \equiv 0$), such that

(i) $v \in C(\mathbb{R}; L^2_w(\mathbb{R}^n))^n$;

(ii) $v(x, t) = 0$ if $(x, t) \notin \Omega \times]0, T[$; in particular $v(\cdot, 0) = v(\cdot, T) \equiv 0$;

(iii) $\dfrac{|v(x, t)|^2}{2} = -\dfrac{n}{2} p(x, t) = \overline{e}(x, t)$ for all $t \in]0, T[$ and almost all $x \in \Omega$;

(iv) $\displaystyle\sup_{0 \le t \le T} \|v(\cdot, t)\|_{H^{-1}(\mathbb{R}^n)} \le \eta$.

Moreover,

(v) $(v, p) = \displaystyle\lim_{k \to \infty} (v_k, p_k)$ in $L^2(dx\, dt)$,

where each (v_k, p_k) is a pair of C^∞ functions with compact support—the classical solution of the Euler equations with a well-chosen forcing $f_k \in C_c^\infty(\mathbb{R}^n \times \mathbb{R}; \mathbb{R}^n)$, $f_k \longrightarrow 0$ in the sense of distributions.

SIXTEEN

Plenty of peace and quiet here! The woods, the gray squirrels, the pond, biking.

And good food! The other day in the dining hall for lunch we had a velvety pumpkin soup, just like at home, a grilled swordfish filet, very tender and well seasoned, a dessert with mulberries and cream that melted in your mouth. . . .

In the afternoon we've only just gotten back to work in our offices when the bell in the clock tower atop Fuld Hall chimes three o'clock: time to go drink tea in the common room and eat the freshly baked cakes that change every day. The madeleines in particular are irresistible, every bit as scrumptious as the ones I used to make for the boys and girls in my dormitory fifteen years ago.

Bread is a real weak point: the crispy French-style baguette is hard to find in Princeton. An even more serious deficiency, as far as products of the highest necessity are concerned, is how scandalously poor the cheese is. The fruity Comté, the delicate Rove, the fragrant Échourgnac, the smooth Brillat-Savarin, the soft Navette, the spicy Olivia, the indestructible Mimolette—I can't find any of them anywhere. My entire family has been suffering since we got here!

Earlier this month I made a lightning visit to the West Coast, to the Mathematical Sciences Research Institute at Berkeley. The MSRI is the world's foremost sponsor of mathematical programs and workshops, welcoming hundreds of visitors every year in addition to a smaller number of research fellows invited for extended stays. It was an emotional moment for me, being back in Berkeley, where I lived for five months in 2004.

Naturally I made a special point of visiting the Cheeseboard, my favorite place in town, a cooperative run on socialist principles (just as you'd expect—this is Berkeley, after all!) offering a selection of cheeses that would put most shops in France to shame.

I loaded up. There was even some Rove, which I knew would make my kids happy. They can't eat enough of it. And when I complained to the people working there about how hard it is to find good cheese in Princeton, they told me to check out Murray's the next time I go to Manhattan. Can't wait!

The French equivalent of the MSRI is the Institut Henri Poincaré in Paris, founded in 1928 with the help of the Rockefeller Foundation and the Rothschild family. Two months ago the governing board of the IHP informed me that I had been selected as the next director of the institute—unanimously, I was told. But I didn't accept at once. I set a number of conditions, and deliberations dragged on and on.

I was first approached about the directorship four months ago. Once the initial surprise wore off, I decided it would be an interesting experience and agreed to be considered as a candidate. I didn't tell my colleagues at ENS-Lyon, fearing they would take it badly. Why should I want to be director of an institute when I'd refused to be director of our laboratory? Why leave Lyon for Paris when I have flourished in Lyon? And who in this day and age really wants to be head of a major scientific organization, weighed down with administrative duties and having constantly to comply with new government regulations and legislative mandates?

How naïve I was to imagine that my candidacy could remain a secret! Not in France . . .

My colleagues in Lyon quickly learned of it—and they were amazed. Why would a mathematician my age seek to be appointed to a position with such burdensome responsibilities? I must be hiding something, they whispered among themselves. There must be some personal secret, some private agenda.

There's no secret. And no agenda beyond a sincere desire to do something new and challenging. But only under the right circumstances! The news wasn't very encouraging. In fact, there wasn't any news at all. Deliberations at the IHP seemed to have gotten bogged down. . . .

Would we be pushing off to Paris, then, or going back to Lyon? Perhaps neither one. Cheese or no cheese, life here is very agreeable, and I have an offer to stay at the IAS for a year, maybe longer if things go well, with a handsome salary and other benefits. And now that Claire's been able to get on with her own research again, this is a good place for her to be too. She's part of a team in the Department of Geosciences at Princeton that's analyzing what may turn out to be the oldest known animal fossils—an extraordinary discovery! The leader of the team is urging her to apply for a postdoc. As it is, by coming with me to Princeton she lost her teaching position in Lyon, and by now it's too late for her to be considered for the next round of faculty assignments.

None of this makes Claire really want to go back. Staying on here would certainly be simpler for her, and more satisfying as well. And so it's difficult to resist the allure of Princeton. To be sure, I can't see myself settling permanently in a place where good bread is so hard to find. . . . But for a few years, why not? And if the IHP can't be bothered to come up with an attractive offer, well, there's nothing I can do about it!

Anyway, I'd been mulling all this over for several weeks and just last night decided to send a letter to France declining the job.

But this morning, when I went to open my email, there it was, a message from the IHP saying that all my conditions had been accepted! Okay to more money, okay to no teaching duties, okay to continued research funding. All of which would have been approved in the United States as a matter of course, but in France it's quite unheard of. Claire was reading the message over my shoulder.

"If they can be counted on to do everything they say they will, you ought to accept."

Exactly what I was thinking. And so it's decided: we will say goodbye to Princeton and go back to France at the end of June!

Now to tell my new colleagues here the news. No doubt some of them will understand and offer their encouragement. (Give it all you've got, Cédric! It's going to be a tremendous experience, etc.) Others will be worried for me. (Cédric, have you really given this enough thought? Running an institution like that will leave you no time for your own work, etc.) One or two, I'm willing to bet, will be terribly upset with me. In any event, my diplomatic skills are going to be tested right away—in the United States rather than France!

In the midst of all this confusion, one thing is certain: nothing is more important right now than the work I'm doing with Clément.

•

The Institut Henri Poincaré ("Home of Mathematics and Theoretical Physics") was founded in 1928 to put an end to the state of isolation in which French science found itself following World War I. Soon it was renowned not only as an institution of scientific training and research but as a cultural forum as well. Einstein lectured on general relativity there, Volterra on the use of mathematical methods in biology. The IHP was home to the first French institute of statistics and the site of the first French computer project. It was also, and not least, a place where artists mingled with scientists. Some of the surrealists found inspiration there, as the photographs and paintings of Man Ray attest.

Later incorporated as a branch of the mathematics faculty of the University of Paris, the IHP was moribund for two decades after 1968, then reestablished in the early 1990s as a department of Université Pierre et Marie Curie (UPMC) and an organ of national scientific policy supported by the Centre National de la Recherche Scientifique (CNRS).

As an integral part of a very large university, the IHP is assured of financial stability during uncertain times and benefits from the expertise of a sizable staff of administrators and technical specialists that a private institution of its size could not afford to maintain. Affiliation with the CNRS provides additional financial and administrative resources, as well as direct access to a national network of scientific research organizations.

The IHP does many things. Above all it serves as a meeting place for mathematicians and physicists from France and around the world. In addition to offering a graduate-level curriculum, it sponsors lectures and tutorials for students and visiting researchers on selected topics, and hosts a great many conferences and seminars every year, welcoming an endless stream of invited speakers. The thirty members of the IHP's governing body, elected in part by national ballot, include representatives of all the major scientific institutions in France; the twelve seats on the scientific council, which is totally independent of the board of directors, are occupied by researchers of the first rank. The Institute's historic location, its outstanding reference library, its commitment to both teaching and research, and its close partnership with learned societies and other associations devoted to the advancement of mathematics and physics all contribute to its great and enduring influence.

[From the mission statement of the Institut Henri Poincaré]

SEVENTEEN

The children are back from school, merrily building a playhouse. The squirrels are prancing on the lawn outside. . . .

I'm on the phone with Clément. He's feeling rather less merry.

"We can resolve some of the problems I mentioned using stratified estimates. But there are still loads of other problems. . . ."

"Well, at least we're making progress."

"I've been studying Alinhac–Gérard, and there's a serious problem with the estimates: you'd need some leeway regarding regularity to get convergence to zero in the regularization term. And it gets worse: regularization could eliminate the bi-exponential convergence of the scheme."

"Damn! I missed that! You're *sure* that Newton's rate of convergence is lost? Well, we'll figure out something."

"And the regularization constants in the analytic are monsters!"

"They're a big problem, absolutely. But we'll find a way to deal with them."

"And then in any case the constants are going to blow up way too fast to be killed off by the convergence in Newton's scheme! Since the background has got to be regularized in order to cope with the error created by the function b, it's the inverse of the time, but there's a constant, and this constant has got to make it possible to control the norms that come from the scattering. . . . Don't forget, these norms grow in the course of the scheme, since we want the losses to be summable on lambda!"

"Okay, okay. I don't know yet how we're going to handle all of this. But I'm sure we'll find a way!"

"Cédric, do you believe—*really* believe—that we'll be able to handle all of this by means of regularization?"

"Yes, of course. These are all technical details. Look at what we've done overall. We've made a hell of a lot of progress. We've figured out how to deal with resonances and the plasma echo, we've worked out the principle of time cheating, we've got good scattering estimates, good norms—we're almost there!"

Clément must think that I'm a pathological optimist, one of those people you're crazy to have anything to do with, the sort of person who goes on hoping when there's no hope left to be had. This latest difficulty does indeed look formidable, I admit. But still I feel sure we'll find a solution. Already three times in the last three weeks we've found ourselves at an impasse, and each time we've found a way around it. It's also true, however, that some obstacles we thought we'd put behind us have come back to block our way in another form. . . . Nonlinear Landau damping is a monster, no doubt about it—with as many heads as the Hydra! Nonetheless, I remain convinced that nothing can stop us. *My heart will conquer without striking a blow.*

•

Date: Mon, 2 Feb 2009 12:40:04 +0100
From: Clement Mouhot <clement.mouhot@ceremade.dauphine.fr>
To: Cedric Villani <Cedric.VILLANI@umpa.ens-lyon.fr>
Subject: Re: aggregate-10

Here are some comments off the top of my head:

- I'm confident for the moment about the norms with two
shifts, I'm looking carefully at the scattering to see if the

estimates I've got are sufficient to express it in terms of norms with two shifts,

- ok for section 5, it fits nicely with the transfer of regularity + gain in decay, that's a really lovely trick! Am I right in supposing that the "gain in decay" part carries forward the "big" phase interval to only one of the functions (becoming an interval between both shifts of the norm with two shifts), in the hope that by applying it to the field created by the density everything will work out smoothly?

- as for section 6, ok for the general idea and the calculations, but (1) I'd rather not try to sum the series in k and l since the coeffs don't seem to be summable (no big deal), (2) to be able to assume that epsilon is small in th. 6.3, it seems to me we've got to make c small as well--is that borne out by what comes later? More comments to follow . . . best, clement

Date: Sun, 8 Feb 2009 23:48:32 -0500
From: Cedric Villani <Cedric.VILLANI@umpa.ens-lyon.fr>
To: Clement Mouhot <clement.mouhot@ceremade.dauphine.fr>
Subject: news

So, two pieces of good news:

- after reading two articles on the plasma echo I realize that this phenomenon is caused by exactly the same "resonances" that were such a headache in section 6. It's all the more astonishing when you consider that they use almost identical notations, with a \tau . . . This makes me even more convinced that the danger identified in section 6 is physically significant, in short, it is a question of determining whether the SELF-CONSISTENT ECHOES in the plasma are going to accumulate and eventually destroy the damping.

–I think I've found the right way to treat the term $\ell = 0$ that I had "provisionally" set aside in section 5 (in σ_0 of Theorem 5.8: it is estimated like the others, but one keeps all the terms, and makes use of the fact that $\| \int f(t,x,v) \, dx \| = O(1)$ in large time (or rather $\|\int \nabla\_v f(t,x,v) \, dx \| = O(1)$). This IS NOT a consequence of our estimate over $f(t,x,v)$ in the gliding norm, it's an additional estimate. For a solution of the free transport, $\int f(t,x,v) \, dx$ is preserved over time, therefore it's perfectly reasonable. When scattering is added, it will no longer be $O(1)$, but $O(t-\tau)$ or something of the sort, and then this has to be killed by the exponential decay in $t-\tau$ that I've kept in the present version of Theorem 5.8.

The modifications that I've made in the version attached here are:

* sections 1 and 2 modified to fully take into account these papers on the plasma echo (I hadn't really understood what the experiment involved, and probably all the math hounds missed the crucial importance of this move, here I believe we're miles ahead of everyone else)

* subsection added at the end of section 4 to make it clear which time norms we're going to be working with; I mention this business of regularization by spatial means, which moreover is equally consistent with the sources cited by Kiessling

* section 5 modified to take into account the treatment of
the term \ell=0

* reference added on the plasma experiment

An IMPORTANT CONSEQUENCE is that in section 8 it will be
necessary not only to propagate the gliding regularity on f,
but also to propagate the uniform (in t) regularity (in v)
on \int f dx.

I haven't made any modif in section 7 but as you must have
guessed, what I've put in section 7.4 "Improvements" is
outdated, I wrote it before realizing that it was the
difference (\lambda \tau+\mu) - (\lambda' \tau'+\mu'), or
something of this ilk, that should really count.

I haven't modified section 8 either but a lot of what I
wrote there concerning the "zero mode" of f_ \tau is
likewise obsolete.

What news on your end? Everything now depends on section 7.

Best

Cedric

Date: Sat, 14 Feb 2009 17:35:28 +0100
From: Clement Mouhot <clement.mouhot@ceremade.dauphine.fr>
To: Cedric Villani <Cedric.VILLANI@umpa.ens-lyon.fr>
Subject: Re: final aggregate-18

So here's version 19 with a complete version of the
statements of theorems 7.1 and 7.3 on scattering in a hybrid

norm with one and two shifts. Apparently the composition theorem with two shifts from section 4 is all we need for the proof (phew!). It looks like it should work but you'll have to check carefully, the version with two shifts is still a bit of a mess. I haven't yet incorporated the Sobolev correction, but surely this point is less of a worry. Otherwise I've modified one detail (also in the theorem with one shift): the estimates of losses on the indices and on the amplitude are now not only uniform, but tend to zero in \tau \to +\infty, as required in section 8. And these losses are in O(t-\tau) for small (t-\tau). I'll get back to work tomorrow, adding the Sobolev correction and completing section 8 so it follows on from section 7.

Best regards, clement

Date: Fri, 20 Feb 2009 18:05:36 +0100
From: Clement Mouhot <clement.mouhot@ceremade.dauphine.fr>
To: Cedric Villani <Cedric.VILLANI@umpa.ens-lyon.fr>
Subject: Re: Draft version 20

Here's version 20 (still in draft), the complete stratified theorem with two shifts. There's now a fundamental problem in connection with theorem 5.9: b can't go to 0 during the Nash–Moser scheme since it corrects an error term due to the scattering itself, which doesn't go to zero since it is associated with the field . . . I'm now scrutinizing theorem 5.9.

best regards,

clement

EIGHTEEN

Princeton
February 27, 2009

A bit of a party atmosphere at the Institute today now that the five-day workshop on geometric partial differential equations is coming to an end. Very fine casting, with many stars—all the invited speakers agreed to participate.

In the seminar room I found a place to stand all the way at the back, behind a large table. Sometimes an audiovisual control board is set up on it, but not today, so I could spread out my notes on top. There's no better place to be. I was lucky to get there before Peter Sarnak, a permanent professor at the Institute who likes it as much as I do. You can always be sure of staying awake, for one thing. If you're sitting in a chair you're more likely to drift off—and you've also got to settle for writing on a small fold-down tablet.

I like to be able to pace back and forth in my stocking feet when I'm listening to a lecture, ideal for stimulating thought.

At the break I rushed outside without bothering to put my shoes back on and ran upstairs to my office. Quick phone call to Clément.

"Clément, did you see my message from yesterday with the new file?"

"You mean the new scheme you got by writing down the characteristic equations? Yeah, I looked at it and I began to do the calculations. Looks like a bear to me."

Monster, beast, bear—these words occur over and over again in our conversations. . . .

"I have a feeling we're going to run into problems with convergence," Clément continued. "I'm also worried about Newton's scheme and the linearization error terms. There's another technical detail, too—you're always going to have scattering from the previous step, and it won't be trivial!"

I was a bit annoyed that my brilliant idea hadn't convinced him.

"Well, we'll see. If it doesn't work, too bad, we'll stick with the present scheme."

"It's pretty wild—we've got more than a hundred pages of proofs by this point and we're still not done yet!! Do you really think we'll ever finish?"

"Patience, patience. We're almost there. . . ."

The intermission in the seminar room was over. I hurried back downstairs to hear the concluding talks.

•

Partial differential equations express relations between the rates of variation of certain quantities as a function of different parameters. PDEs constitute one of the most dynamic and varied domains of the mathematical sciences, defying all attempts at unification. They are found in every phenomenon studied by the physics of continuous systems, involving all states of matter (gases, fluids, solids, plasmas) and all physical theories (classical, relativistic, quantum, and so on).

But partial differential equations also lurk behind many geometric problems. Geometric PDEs, as they are called, make it possible to deform geometric objects in accordance with well-established laws. The application of methods of analysis to problems in other fields of mathematics is an example of the sort of cross-fertilization that became increasingly common in the course of the twentieth century.

The February 2009 workshop at the IAS addressed three principal themes: conformal geometries (involving transformations on a space that distort distances but preserve angles), optimal transport (the movement of mass from an initial defined configuration to a final defined configuration while expending the least energy possible), and free-boundary problems (the form of the boundary that separates two states of matter or two materials). Three topics that touch on geometry and analysis in addition to physics.

In the 1950s, John Nash disrupted the balance between geometry and analysis when he discovered that the abstract geometric problem of isometric embedding could be solved by the fine "peeling" of partial differential equations.

A few years ago, Grigori Perelman solved Poincaré's conjecture by using a geometric PDE known as the Ricci flow, invented by Richard Hamilton. Once again the analytic solution of an emblematic problem in geometry shattered the status quo and created an unprecedented interest in exploring other applications of geometric PDEs. The shock waves from Perelman's bomb were felt throughout the world of mathematics—an echo of the one Nash had set off fifty years earlier.

NINETEEN

Princeton
March 1, 2009

I read the message that had just appeared on my computer screen, and then read it again. Couldn't believe my eyes.

Clément's come up with a new plan? He wants to give up on regularization? Wants to forget about making up for the loss of regularity encoded in the time interval?

Where did all this come from? For several months now we've been trying to make a Newton scheme work with regularization, as in Nash–Moser—and now Clément is telling me that we need to do a Newton scheme without regularization? And that we've got to estimate along the trajectories, while preserving the initial time and the final time, with *two* different times??

Well, maybe he's right, who knows? Cédric, you've got to start paying attention, the young guys are brilliant. If you don't watch out, they're going to leave you in the dust!

Okay, there's nothing you can do about it, the next generation always ends up winning . . . but . . . *already*?

Save the sniveling for later. First thing, you've got to try to understand what he's getting at. What does this whole business of estimating really amount to, when you get right down to it? Why should it be necessary to preserve the memory of the initial time?

In the end, Clément and I will be able to share the credit for the major innovations of our work more or less equally: I came up with the norms, the deflection estimates, the decay in large time, and the echoes; he came up with the time cheating, the stratification of errors, the dual time estimates, and now the idea of dispensing with regularization. And then there's the idea of gliding norms, a product of one of our joint working sessions; not really sure whose idea that was. To say nothing, of course, of hundreds of little tricks . . .

Perhaps it wasn't such a bad thing after all that an ocean and a few time zones came between us in the middle of the project: for a couple of months each of us has been forced to concentrate on his own strategy without having to listen to any opposing arguments. It's now become clear, however, that our separate points of view will have to be reconciled somehow.

If Clément is right, the last great conceptual obstacle has just been overcome. On this first day of March our undertaking has entered into a new phase, less fun, but also more secure. The overall plan is in place, the period of free-ranging, open-ended exploration is over. Now we've got to consolidate, reinforce, verify, verify, verify. . . . The moment has come for us to deploy the full firepower of our analytical skills!

Tomorrow I'm taking care of the kids; there's no school on account of the snowstorm. But come Tuesday, the final push begins. One way or another the Problem simply has *got* to be tamed, even if it means going without sleep. I'm going to take Landau with me everywhere—in the woods, on the beach, even to bed. Time now for *him* to watch out!

Long afterward, Clément confessed to me that he had decided to bail earlier that weekend. On Saturday morning, February 28, he began to compose an ominous message: "All hope is lost . . . the technical hurdles are insurmountable . . . can't see any way forward . . . I give up." But just as he was about to send it, he hesitated. He wanted to find the right words to convince me, but also to console me. So he saved his message in the draft folder. Going back to it that evening, armed with pencil and paper in order to make a list of all the paths we had explored and all the dead ends they had led us into, he saw, to his amazement, the right way to proceed opening up before him. The next morning, having gotten up at six o'clock after a few hours of fitful sleep, he wrote out everything again so he'd have a clean copy of the key idea that might just save us, and then finished his message to me.

That day, Sunday, March 1, we came within a whisker of abandoning our dream. Several months of work very nearly disappeared—at best, filed away in a drawer; at worst, gone up in smoke. On the other side of the Atlantic, however, I had no idea that we had come within an inch of catastrophe. All I sensed was the enthusiasm emanating from Clément's message.

During the month of February I exchanged a good one hundred emails with Clément. In March, more than two hundred!

•

Date: Sun, 1 Mar 2009 19:28:25 +0100

From: Clement Mouhot <clement.mouhot@ceremade.dauphine.fr>

To: Cedric Villani <Cedric.VILLANI@umpa.ens-lyon.fr>

Subject: Re: aggregate-27

Might be some hope if we go at it from another angle: not to regularize but to try to propagate the norm to a shift that's needed at each step of the scheme, only along the characteristics of the preceding step. In order, then, we would estimate at rank n (I don't write down the summable losses on lambda and mu each time):

1) F norm of the density \rho _ n with lambda index t +mu

2) Z norm of the distrib h _ n with lambda coefficient, mu and t

3) C norm of the spatial average <h _ n> with lambda index
4) Z norm at time tau with a shift -bt/(1+b) along the
(complete) characteristics S _ {t,tau} of order n-1.
Differentiating with respect to tau we obtain an equation on
H _ tau := h^n _ tau \circ S _ {t,tau} ^{n-1}
of the type (I don't include any possible minus signs)
\partial _ tau H = (F[h^n] \cdot \nabla f^{n-1})
\circ S _ {t,tau} ^{n-1} + (F[h^{n-1}] \cdot \nabla h^{n-1})
\circ S _ {t,tau} ^{n-1}
So basically in this equation there's no longer any field
at all and we treat the whole right-hand term as a source
term, by using the bounds on density in point 1) above:
The Z norm is estimated with the b shift: as for the
density we treat the error committed on account of the
characteristics by means of this shift (since the norm is
projected on x) and for the other terms we use the
recurrence assumption from the preceding point to bound
the existing norms.
5) Now we need to have a bound (in shifted norm) on
f^n \circ S _ {t,tau} ^n (with well-defined characteristics n),
by using the bound of the recurrence assumption
(in shifted norm) on f^{n-1} \circ S _ {t,tau} ^{n-1}.
Thanks to 4) above, by addition we get a limit on
f^n \circ S _ {t,tau} ^{n-1}. Then we've got to exploit
the possibility of bounding f^n \circ S _ {t,tau} ^n
(characteristics of step n) as a function of f^n \circ S _
{t,tau} ^{n-1} (characteristics of step n-1) modulo a loss,
summable as n goes to infinity.

The general idea may be summarized as follows:

- To estimate the density there isn't any choice, we've got
to have characteristics and a shifted norm (with a shift of

order 1) on the distribution of the preceding step, along
the characteristics of the preceding step,

– But once you've got the bound on the characteristics, you
can work along the characteristics and in shifted norm,
since when they're projected onto the density, the two
phenomena cancel each other.

There's one thing I omitted to mention, the gradient in v on
the background, which doesn't commute with the composition
by the characteristics, but one might hope to have something
like shifted norm of (\nabla _ v f^{n-1}) \circ S _ {t,tau}
^{n-1} smaller than constant times shifted norm of \nabla v
(f^{n-1} \circ S _ {t,tau} ^{n-1}) . . .

If you're around we can talk about it on the phone I'm at
home for another hour: I think that this accords with your
scheme for the most part, with the difference that it
fundamentally distinguishes between two steps and only later
looks at things along characteristics.

Best regards, Clement

Date: Mon, 2 Mar 2009 12:34:51 +0100
From: Clement Mouhot <clement.mouhot@ceremade.dauphine.fr>
To: Cedric Villani <Cedric.VILLANI@umpa.ens-lyon.fr>
Subject: Version 29

So here's version 29, in which I've really tried to implement
the strategy I spoke to you about yesterday: it's in section
9 on linear stability which I've entirely rewritten, and

subsections 11.5 and 11.6 of the section on the Newton scheme where I've sketched the convergence study. Unless I'm hugely mistaken, I really have the impression that we're nearing our goal!!

TWENTY

Lunch in the dining hall is always a treat. Delicious food and lively conversation about mathematics—and other mathematicians!

Today Peter Sarnak was sitting across from me, a great talker. I got him started on the subject of Paul Cohen, who proved the undecidability of the continuum hypothesis before turning his attention to other topics. Peter had left his native South Africa as a young man to study with Cohen at Stanford, lured away from home by the thrill of mathematical exploration. With a few quick strokes Peter drew a memorable picture of his teacher, and emphasized Cohen's insistence on solving problems ex nihilo, without relying on the work of others.

Peter Sarnak

"Cohen didn't believe in incremental mathematics."

"Incremental?"

"Yes, he thought that mathematics progresses by sudden leaps. You and me, like everyone else, nudge it forward by improving on the work of others, but not Cohen! If you told him you were improving something he'd send you packing. He believed in revolutions and nothing else!"

Always a pleasure listening to Peter. Also at our table was the young Israeli mathematician Emanuel Milman, a rising star of convex geometry whose office is right next to mine. Emanuel is the son, nephew, and grandson of mathematicians. Recently he became a father. Of a future mathematician? Certainly he speaks with no less delight about his wonderful child than about his own work.

Seated next to Emanuel was Sergiu Klainerman, who fled Communist Romania in the 1970s. Sergiu achieved worldwide fame when, together with the phenomenal Greek mathematician Demetri Christodoulou, he established a fundamental result of general relativity in an epic five-hundred-page proof. I very much enjoy discussing mathematics, politics, and ecology with Sergiu, subjects on which we often disagree. . . .

By no means the least of the reasons why the conversation at our table was so animated was the legendary verve of Joel Lebowitz, who, despite his advancing years, has still got way more energy than people half his age. Joel is interested in everything, wants to know everything, and if you get him talking about his favorite subject, statistical physics, he can go on forever.

I took advantage of Joel's presence to ask him to tell Emanuel about the difficulties involved in explaining phase transitions for a hard sphere gas. Simple to state, of fundamental importance, and yet the problem has baffled statistical physicists for more than fifty years now. I mean, really, here we are today, in 2009, and *still* we haven't managed to penetrate the mysteries of a state change—why a liquid is transformed into a gas when it is heated, why it is trans-

Joel Lebowitz

formed into a solid when it is cooled! Who knows, a really smart young guy like Emanuel just might have a new idea. . . .

No sooner had the lunch break come to an end than all my nagging worries came flooding back. There are still some administrative questions to be settled with the Institut Poincaré, or rather the question of my association with ENS-Lyon, which I would like to maintain while at the same time fulfilling my duties as director in Paris. At the lab in Lyon my great pal Alice Guionnet has my back while I'm gone, but everything is so complicated. . . . Then there's the series of talks I've got to prepare. And then there's the most important thing of all, the Landau damping proof, which still doesn't want to come together! Over the past ten days Clément and I have drafted ten new versions of our article, the latest one is number 36 and runs to one hundred thirty pages. We've spotted and corrected a number of errors and added a very instructive section of counterexamples. Another one of my colleagues in Lyon, Francis Filbet, has provided us with marvelous computer-generated images of Landau damping as well. But there's still so much to do! A voice in my head echoes softly: *We've got to refine the estimates on the characteristics and get the supremum inside the norm, concentrate on the @!*# Coulomb* *interactions, add a Sobolev regularity correction index more or less everywhere (seven indices—porca* miseria!*), keep the exponential stratification along the Newton scheme, get the huge recurrence going round and round. . . .*

My big mistake was to let the indefatigable Joel drag me into a working session with another French mathematician. Before long I found myself overwhelmed by a sense of desperation. There are so many things I should be focusing on; for the last several days I've been working until two in the morning . . . in the haze of my après-lunch torpor I was scarcely able even to collect my thoughts. Impossible to say no to Joel—but seeing that the session was going to drag on and on, I broke down and resorted to a shameful piece of subterfuge: Sorry, I've got to go now, got to pick up the kids from school (when today that was actually their mother's job). I hid outside until the two of them had gone down the hall to work in another room, then crept back into my office, lay down on the bare floor, and went to sleep so that my poor overtaxed brain could have a moment's rest.

Soon I was awake again and back to work at once.

•

Paul Cohen, briefly Nash's younger colleague and ambitious rival at MIT in the late 1950s, was one of the most creative minds of the twentieth century. His greatest claim to fame is the solution of the continuum hypothesis (also known as the problem of the intermediate cardinal). This enigma, the first of the twenty-three outstanding problems stated by Hilbert in 1900, was still considered one of the most important challenges in mathematics more than a half century later. Anyone who could solve it was obviously in line for a Fields Medal, duly awarded to Cohen in 1966.

To understand the continuum hypothesis, it will be helpful to keep a few things in mind. The whole numbers (1, 2, 3, 4 . . .) are, of course, infinite in number. The fractional numbers (1/2, 3/5, 4/27, 53417843/14366532, . . .) are infinite as well. It seems as though fractional numbers must be more numer-

*ous than whole numbers, but this is an illusion: one can enumerate the fractions,
for example*

1, 1/2, 2/1 [= 2], 1/3, 3, 1/4, 2/3, 3/2, 4, 1/5, 5, 1/6, 2/5, 3/4,
4/3, 5/2, 6, . . .

*by gradually increasing the sum (numerator + denominator)—as Ivar Ekeland
explains so well in his delightful children's story,* The Cat in Numberland.
*Therefore there aren't more fractional numbers than whole numbers, there are
exactly as many of the one kind as of the other.*

*But if we consider the so-called real numbers—those numbers that can be writ-
ten down only with an infinite number of decimal places (which are also the limits of
fractional numbers)—then, thanks to a magnificent argument devised by Cantor,
we see that there are many more reals. Indeed, there are uncountably many of them.*

*We therefore have an infinite quantity of whole numbers (or integers), and a
still greater infinite quantity of real numbers. So, is there an infinity that is both
greater than the infinity of the integers and smaller than the infinity of the reals?
Mathematical logicians worked on the problem for decades, some trying to show
that yes, this intermediate infinity does exist; others, to the contrary, that no, it
doesn't exist. But to no avail.*

*Paul Cohen wasn't a logician by training, but he believed in his own brain-
power. To everyone's amazement he succeeded in showing that the answer is neither yes nor no: there does exist a mathematical world that has an intermediate
infinity, yet there also exists a mathematical world that does not have one—and
it is up to us to choose which world we want. Either one will be correct, if it is the
one we want. Determining which world is the more natural of the two remains an
open problem in set theory still today.*

•

*Joel Lebowitz is the high priest of statistical physics, the branch of science that seeks
to discover the properties of systems consisting of very large numbers of particles.
Gases made up of billions of billions of molecules, biological populations made up of
millions of individuals, galaxies made up of hundreds of billions of stars, crystalline
lattices made up of billions of billions of atoms—statistical physics covers a lot of*

territory! And for almost sixty years Joel has put his inexhaustible energy in the service of an unequaled passion for the subject, tirelessly working with his fellow mathematicians and physicists. The semiannual series of conferences he inaugurated in 1958 is probably the oldest meeting organized by an active scientist today. I doubt any other meeting has welcomed more participants over the years.

Born in Czechoslovakia nearly eighty years ago, Joel has had such a full life, full of good and bad memories both. A number is tattooed on his forearm; he never speaks of it. No matter what the occasion, Joel is the first one to laugh, the first one to hoist a glass, and, of course, the first one to talk about statistical physics—as though it were a vast musical composition whose every theme he has set out to explore, playing in every key. The amount of energy a person has ought to be measured in "milli-Joels," thousandths of Joel, as one of our colleagues wittily suggested: a thousandth of Joel's energy is still a good amount. Maybe even "pico-Joels," come to think of it!

•

Date: Mon, 9 Mar 2009 21:42:10 +0100
From: Francis FILBET <filbet@math.univ.-lyon1.fr>
To: Cedric Villani <Cedric.VILLANI@umpa.ens-lyon.fr>
Cc: Clement Mouhot <clement.mouhot@ceremade.dauphine.fr>

Hello

Here's what I was able to work up over the weekend. Short films, no big deal (not on a level with Desplechin): in the part about numerical simulations of charged particle systems.

http://math.univ-lyon1.fr/~filbet/publication.html

This is the plasma case. I haven't changed the sign for the gravity case yet I'm very surprised by what you say. I think that a neutralizing background is needed to keep a periodic potential i.e. $\int_0^L E(t,x)dx = 0$ when one has periodic boundary conds

Date: Mon, 9 Mar 2009 22:11:10 -0500
From: Cedric Villani <Cedric.VILLANI@umpa.ens-lyon.fr>
To: Francis FILBET <filbet@math.univ.-lyon1.fr>
Cc: Clement Mouhot <clement.mouhot@ceremade.dauphine.fr>

The images are superb! It's really moving to see equations on which one has worked "in the abstract" brought to life, made real. . . .
Cedric

TWENTY-ONE

Princeton
March 13, 2009

Closed the door to the kids' room a moment ago. My daughter was still chuckling about the adventures of Goofy, the hero of tonight's imaginary story. *Sleep, little wonders, tomorrow day will dawn.*

Claire's in bed, too—last chance to brush up on her Japanese before heading off bright and early tomorrow on a field trip with her research team. Just the right moment for me to get some work done. Made a pot of tea and spread out my drafts on the floor. Still a mountain of technical problems, which Clément and I go on climbing one problem at a time.

Working on section 9 at the moment, the longest section of the proof. There's this damn control on the zero mode, I knew it was going to drive me crazy. And I've got to present the result in ten days! Ten short days to get the whole thing to hang together.

•

Date: Fri, 13 Mar 2009 21:18:58 -0500
From: Cedric Villani <Cedric.VILLANI@umpa.ens-lyon.fr>
To: Clement Mouhot <clement.mouhot@ceremade.dauphine.fr>
Subject: 38!

Version 38 attached. The modifs:

- 2-3 typos corrected here and there as you can see with diff if you need to

- section 9 is now complete modulo a certain number of formulas, this is the time to summon our courage and see all the calculations through to the end! It's a rather beautiful thing to behold since all the pieces fit together and lead on to the result. The organization of this section justifies the paper's overall structure a posteriori (in particular putting the characteristics at the beginning). With a little more tinkering this section ought to be in good shape, and then we'll be ready to decide about constants (Hello calculations!)

- I've cut out most of the old commentary, particularly the remarks concerning regularization.

- But there are still two holes in the way the spatial averages are handled!

* the first one has to do with the need to stratify the estimates on $< \nabla h^k \circ \Om^n >$ (subsection 9.4). This is tricky, as I explain in the file, we can't rely on recurrence, and we can't rely on regularity since $\Om^n$ is highly irregular. The only solution I see is to use the additional Sobolev regularity of the characteristics, which is propagated uniformly over n. Be careful, it's velocity regularity we need, but that should be OK, Sobolev regularity for the force entails regularity for all variables. We've got to gain exactly one derivative, which means that Coulomb's probably critical here as well. . . . .

* the second one is the treatment of the zero mode in the estimates in section 6, where for the moment it doesn't work

(constants too large to go on checking the stability criterion). I'm fairly optimistic, counting on being able to recycle my old idea of using the change in the scattering variables, and DIRECT estimates on the characteristics. The first time I tried it I didn't have the right orders of magnitude in mind, we hadn't done the stratification yet, in other words we were much less well prepared.

I suggest the following division of labor: first you see about making section 9 converge, forgetting about the two holes above; then you look for a way to plug the first hole. In the meantime I'll get to work on the second hole. I'm not planning to make any changes to the tex file in the next few days.

As for the Coulomb case: we'll see later, I think that plugging the holes has to take priority. . . .

This week is going to be a little hard for me because I'll be taking care of the kids by myself, and on top of this we've got guests at the lab. But it's pretty much the final sprint.

Best
Cedric

TWENTY-TWO

Princeton
Night of March 15–16, 2009

Sitting on the floor, surrounded by sheets of scribbled notes strewn all over the carpet, I write and type for hours in a state of feverish excitement. . . .

Made a point of not doing any mathematics during the day. Took the kids along to Sunday brunch at Alice Chang's house, many famous names in attendance. Alice teaches at the university, a renowned specialist in geometric analysis; a few years ago she was a plenary speaker at the International Congress of Mathematicians in Beijing. It was Alice who invited me to give a series of lectures as part of a program she had organized at Princeton this spring on differential geometry and geometric analysis.

This morning at brunch we talked about a little bit of everything, including the famous Shanghai ranking, the list of the world's top universities that French politicians and journalists are so fond of citing. When I raised the subject with Alice I wondered how she would react, since she is both Chinese by birth and a member of one of the most celebrated mathematics departments in the world. I thought she might be proud of the reputation this classification has acquired outside her native land. Boy, was I ever wrong!

"Cédric, what's the Shanghai ranking?"

Alice Chang

When I explained what it was all about, she looked at me as if I were pulling her leg. Cédric, I don't follow—being on this *Chinese* ranking is considered very prestigious *in France*?? (Are you sure you don't have it backward, my dear?) How I'd love to introduce Alice to some politicians I know back home. . . .

It was only much later, well into the evening, in fact, after the children had gone to sleep, that I finally got down to work. And then the miracle occurred. Everything seemed to fit together as if by magic!

Trembling all over, I write out by hand and then type into the computer file the last six or seven pages—all the loose ends of the proof are tied up at last, at least for interactions more regular than the Coulomb interaction.

At 2:30 I go to bed. My head feels like it's going to explode. I stay awake for a long, long time, eyes wide open.

At 3:30 I finally fall asleep.

At 4:00 my son comes in to wake me up, he's wet his bed. It's been years since that's happened—had to happen tonight, of all nights. . . .

That's life. Onward. I get up, change the sheets, remake the bed. The whole routine.

There are times when everything conspires to prevent you from sleeping. So be it!

•

Every mathematician worthy of the name has experienced, if only rarely, the state of lucid exaltation in which one thought succeeds another as if miraculously. . . . Unlike sexual pleasure, this feeling may last for hours at a time, even for days.

[André Weil]

TWENTY-THREE

Princeton
March 22, 2009

In the end it turned out that my solution didn't quite work. It took me almost a week to convince myself of this. The better part of the proof survives intact, but the abominable zero mode continues to taunt us. Still, we came close!

Clément is in Taiwan, giving the first public talks about our work. He pondered my ideas, combined them with his own, and expressed the result in his own way. Then I reworked it and expressed it in mine.

The new version is much simpler than my first attempt—and it works! We've been slaving away at the proof for exactly a year now, and for the first time it really seems to hang together!

The timing couldn't be better: I'm due to announce the result in Princeton two days from now. . . .

•

Date: Sun, 22 Mar 2009 12:04:36 +0800
From: Clement Mouhot <clement.mouhot@ceremade.dauphine.fr>
To: Cedric Villani <Cedric.VILLANI@umpa.ens-lyon.fr>
Subject: Re: finishing touches

Okay, I think I finally understand what you had in mind for the spatial average!! And I think it has to be combined with

the idea I mentioned to you on the phone (in fact the two
are complementary), here's the plan:

(1) I think that the calculation you had in mind, using the
best stratified regularity on the background, is the
calculation at the beginning of subsection 6 (pages 65–66):
in this case (no scattering), one can in fact use the margin
of regularity on the background "for free" to create growth
(independently of the level of regularity on the force
field).

(2) It's a question then of reducing to this case by means
of the idea I mentioned on the phone (the "remainder" that
we talked about isn't trivial, it's got to be treated by (1)):

a. we replace $F[h^{n+1}] \circ \Omega^n _ {t,\tau} \circ S^0
_ {\tau,t}$ by $F[h^{n+1}] \circ S^0 \_ {\tau,t}$, the remainder
has the right time decay thanks to the estimates on
$\Omega^n - Id$

» therefore we're left with

\int _ 0 ^t \int _ v F[h^{n+1}] \cdot < ((\nabla _ v f^n)
\circ \Omega^n) > (x-v(t-\tau),v) \, d\tau \, dv

b. now we make the key move, changing the variable to
replace \Omega^n by \Omega^k in \nabla _ v f^n (for any k
between 1 and n): the pb we used to have applying \Lambda no
longer exists since we no longer compose \Omega^n X with
(\Omega^n)^{-1} \Omega^k, now we have only (\Omega^n)^{-1}
\Omega^k, which we already have estimates on.

c. once again we get rid of the application (\Omega^n)^{-1}
\Omega^k which has been transferred to F[h^{n+1}] by the same

trick as in step a., which creates a nice new remainder term
that steadily diminishes over time,

» therefore we're left with

\sum _ {k=1} ^n \int _ 0 ^t \int _ v F[h^{n+1}] \cdot <
((\nabla _ v h^k) \circ \Omega^k) (x-v(t-\tau),v) >
\, d\tau \, dv

d. only now do we invert the gradient in v with composition
by scattering:

< (\nabla _ v f^n) \circ \Omega^k > = \nabla _ v (< f^n
\circ \Omega^k >)
+ remainder with steady decay into \tau

» therefore we're left with

\sum _ {k=1} ^n \int _ 0 ^t \int _ v F[h^{n+1}] (x-v(t-\tau),v)
\, \nabla _ v U _ k (v) \, d\tau \, dv

with functions U _ k (v) of regularity \lambda _ k, \mu _ k.

e. At this level we finally apply calculation (1) on pages
65-66 for each k, which ought to give a uniform stratified
estimate.

Tell me what you think and if you get the same results for
the calculations . . .

Best regards, Clement

TWENTY-FOUR

The first of my three seminar talks at Princeton. Before a distinguished audience of tough minded mathematical physicists, none of them tougher than Elliott Lieb, cordial but implacable.

Clément is still in Taipei. A time difference of thirteen hours— very nearly the optimal difference for efficient collaboration at a distance! With the added advantage that this way we split up the world: he spreads the good word in Asia, I do the same in the United States.

This time I was ready to take the plunge. It wouldn't be anything like my wobbly performance at Rutgers back in January, I was sure of that: the proof is 90% correct or better, and all the major aspects of the problem have been identified. I was confident, prepared to submit to questioning and to explain the argument.

While the results made quite an impression, Elliott wasn't convinced by the assumption of periodic boundary conditions, which seemed to him wholly unwarranted.

"If it isn't true in the space as a whole, it's meaningless!"

"Elliott, in the space as a whole there are counterexamples. There's no choice but to set boundaries!"

"Yes, but the result has to be independent of boundaries— otherwise it's not physics!"

"Elliott, Landau himself used boundary conditions, and he

showed that the result depended very strongly on boundaries. You don't mean that he wasn't a physicist, do you?"

"But it's meaningless!"

Elliott had gotten up on his high horse. And he wasn't my only critic. Greg Hammett, a physicist at the Princeton Plasma Physics Laboratory (the PPPL, as everyone calls it), didn't much like my assumption of stability in the case of plasmas. Too strong, he felt, to be realistic.

The triumphal reception I was hoping for turned out to be anything but. A real letdown!

•

Elliott Lieb is one of the most famous and formidable figures in mathematical physics today. A professor of both mathematics and physics at Princeton, he has devoted a part of his life to finding an explanation for the stability of matter: What is it that forces atoms to come together rather than keep their distance, standing apart from one another in serene isolation? Why is it that we are physically coherent creatures? Why don't we just melt away into the surrounding universe? Freeman Dyson, one of the twentieth century's greatest physicists (and

Elliott Lieb

now an emeritus professor at the IAS), was the first to pose this problem in mathematical terms. In clearing the way for future investigation, Dyson inspired a younger generation of researchers to carry on the quest.

Elliott gave himself up to it completely. He looked everywhere, seeking a solution in physics, in analysis, in the calculation of energies. Over a long career he has trained many other scientists, created influential schools of thought, and published spectacular proofs that have changed the face of mathematical analysis.

To Elliott's way of thinking, nothing beats a good inequality in trying to understand a problem. An inequality expresses the domination of one term in an equation by another, of one force by another, of one entity by another. Not only has Elliott profoundly improved a number of celebrated statements, including Hardy–Littlewood–Sobolev inequalities, Young inequalities, and Hausdorff–Young inequalities, he has given his own name to two more groups of fundamental relations: Lieb–Thirring inequalities and Brascamp–Lieb inequalities, used today by scientists throughout the world.

Now almost eighty years young, Elliott is still as active as ever. His trim figure testifies to his physical fitness; his barbed wit, dreaded by all, to his mental sharpness. His eyes light up when he speaks of Japan, inequalities, and gourmet cooking (as it happens, the Japanese word for haute cuisine, kaiseki, *also means mathematical analysis).*

TWENTY-FIVE

The first day of April—the day of fishes and fools!

This afternoon watched an episode of *Lady Oscar* with Claire and the children. Marie-Antoinette, Axel de Fersen, and Oscar de Jarjayes spinning round in a whirl of fine phrases and noble sentiments amid the lengthening shadows of the French Revolution.

And this evening, before going to sleep, we watched a YouTube video of Gribouille singing "Le Marin et la Rose." Simply marvelous! There's some great stuff on the Internet.

During the past week I've learned so much from lecturing on Landau damping.

After my first talk, once his irritation had subsided, Elliott shared some valuable insights into the conceptual difficulties of the periodic Coulomb model.

At the second talk I laid out the main physical idea of the proof. Elliott very much appreciated the mixture of mathematics and physics; he seemed not only engaged but genuinely supportive.

By the time of the third talk I'd come up with an answer to Hammett's objection, and I was able to formulate almost optimal assumptions regarding the stability condition and perturbation length.

I'd taken a risk, presenting completely new results that were still only half-baked, but the gamble paid off: their criticisms enabled me to make much faster progress than I could have otherwise! Once again I had to put myself in a vulnerable position in order to become stronger.

And . . . the connection with KAM finally became clear to me!

The ability to detect hidden connections between different areas of mathematics is what has made my reputation. These connections are invaluable! It's a bit like a game of Ping-Pong: every discovery you make on one side helps you discover something new on the other. The connections make it possible to see more of the landscape on both sides.

My first important result, with the Italian mathematician Giuseppe Toscani, came in 1997, when I was twenty-four years old: the unsuspected link between Boltzmann entropy production, the Fokker–Planck equation, and entropy production for plasmas.

The next one came eighteen months later, with my German coauthor Felix Otto: the hidden link between the logarithmic Sobolev inequality and Talagrand's concentration inequality. Two other proofs have been proposed in the years since. . . . This is how I got started exploring the field of optimal transport. Thanks to our paper I was invited to give a graduate-level course at Georgia Tech, which in turn gave birth to my first book.

During my thesis defense in 1998, Yves Meyer marveled at the "miraculous" relations I had brought to light. "Twenty years ago people would have laughed at your work. No one believed in miracles then!" But *I* believe in miracles—and I shall uncover more of them.

In my thesis I recognized four spiritual fathers: my thesis director, Pierre-Louis Lions; my tutor, Yann Brenier; and Eric Carlen and Michel Ledoux, whose works opened up the fascinating world of inequalities to me. In addition to the joint influence of these four teachers, I incorporated other elements and created my own mathematical style, which then evolved—as it pleased chance to bring me into contact with new friends and new ideas.

Three years after my defense, with my longtime collaborator

Laurent Desvillettes, I discovered an apparently improbable link between Korn's inequality in elasticity theory and the production of entropy for the Boltzmann.

After that I developed the theory of hypocoercivity, based on a new analogy between problems associated on the one hand with regularization, and on the other with convergence to equilibrium, for degenerate dissipative partial differential equations.

There was also the hidden link between optimal transport and Sobolev inequalities, which I had detected earlier with Dario Cordero-Erausquin and Bruno Nazaret—a connection that astounded many analysts who thought they really understood these inequalities!

In 2004, as a visiting research professor in the Miller Institute at Berkeley, I had the good fortune to meet another future coauthor, the American mathematician John Lott, then a guest of the MSRI. Together we showed that insights from the study of optimal transport in economics could be used to tackle various problems in non-smooth non-Euclidean geometry, which together make up the problem of so-called synthetic Ricci curvature. The theory that came out of our collaboration, now called the Lott–Sturm–Villani theory, has had the effect of breaking down a few more barriers between analysis and geometry.

In 2007, once again sensing the presence of a preexisting harmony, I managed to demonstrate a strong relationship between the geometry of the tangent cut locus and the curvature conditions necessary for optimal transport regularity—another connection that seemed to come from nowhere, which I proved with Grégoire Loeper.

Each time, a personal encounter set everything in motion. It was as though I had acted as a catalyst! But I also firmly believe in the importance of searching for preexisting harmonies—after all, Newton, Kepler, and so many others have already shown us the way. Everywhere you look, the world is filled with unsuspected connections!

Nope, didn't no one ever suppose
They'd anything at all in common
Him, the sailor who was in Formosa
And her, who was the rose of Dublin

And only a finger on the lips . . .

Nor did anyone ever imagine for a moment that Landau damping and Kolmogorov's theorem had the slightest thing in common.

Except for Étienne Ghys. Tricked, perhaps bewitched by some mischievous sprite, Étienne had somehow divined that there was a connection between the two. Almost one year to the day after our conversation back in Lyon last April, I've finally figured out what it is!

Hmmm . . . a loss of regularity in a perturbative context, due to resonance phenomena, is made up for by a Newton scheme exploiting the completely integrable character of the system that is disturbed. . . .

Amazing that this idea ever occurred to me at all! Who would ever have imagined that something so weird could be real? Landau damping, to begin with—who would ever have believed that it's fundamentally a question of regularity?

•

THE SAILOR AND THE ROSE

Once upon a time there was a rose
Aye, a rose there was and a sailor
The sailor, he was in Formosa
The rose, she was a Dubliner

Didn't never see each other, nope,
Far too great a distance between them
Him, he never left his sailing boat
Her, she never left her fair garden

Over the chaste rose, high in the sky,
Chased swirls of birds in swooping rings
And after them waves of clouds swam by
And after them waves of suns and springs

Over the fickle sailor passed shrouds
Of dreams, each one the same as the ones
Before, dreams of springs and dreams of clouds
Birds of the mind and imagined suns

The sailor, he died in September
And the rose, on the very same day
Wilted in a room free from slumber
Where wept a girl love had thrown away

Nope, didn't no one ever suppose
They'd anything at all in common
Him, the sailor who was in Formosa
And her, who was the rose of Dublin

And only a finger on the lips
Lovely as only lightning can be
Now the sun draws near to its eclipse
An angel casts petals down upon the sea

TWENTY-SIX

Version 55. The tedious process of rereading and fine-tuning. Then, suddenly, a new hole opens up.

Hopping mad, I've just about had it.

Had it up to here *with this whole business! Before it was the nonlinear part. Now it's the linear part that seemed to be under control and then came apart at the seams!*

We've already announced our result more or less everywhere: last week I gave a presentation in New York, tomorrow Clément's doing the same in Nice. There's no excuse for the slightest error at this point—the thing has to be completely correct!!

But there's no getting around it, there *is* a problem. Somehow this damn Theorem 7.4 has got to be fixed. . . .

The children are asleep, Claire's away again. Working in front of the big picture window, looking out into the dark night. The hours go by. Sitting on the sofa, lying on the sofa, kneeling in front of the sofa, I turn my bag of tricks inside out, scribbling away, page after page. To no avail.

I go to bed at four o'clock in the morning in a state close to despair.

•

Date: Mon, 6 Apr 2009 20:03:45 +0200
From: Clement Mouhot <clement.mouhot@ceremade.dauphine.fr>

To: Cedric Villani <Cedric.VILLANI@umpa.ens-lyon.fr>
Subject: Landau version 51

I'm sending you what I've got, after going over everything
line by line for 120 pages. I'm exhausted, going to take
the night off. Here's version 51, which should incorporate
(having now carefully reviewed your messages) all your
modifs and email requests (figures, remarks, dependence of
constants . . .), as well as your rewritten section 10 (from
the last version you sent me, version 50) and the new
section 12.

For my part, I've reread everything up through section 9 (in
other words up through page 118). There are a fair number of
NdCMs that you'll have to look at, but also a bunch of minor
corrections that don't seem to me to require any discussion.
Only two of my comments raise concerns about the proofs (but
in neither case do they call the result into question):
section 7 page 100 and section 9 page 116.

Here's what I suggest we do in the next couple days: how
about if you work from this version 51 and go over sections
1 to 9, looking at all my comments and deleting each one in
turn once you've dealt with it, in what let's call version
51-cv, and I'll carefully reread sections 10-11-12-13-14? (can
we both aim to finish up by tomorrow evening or Wednesday
morning?)

Best regards, Clement

TWENTY-SEVEN

Uhhhhh . . . man, is it hard to wake up! Finally, with the greatest difficulty, I manage to sit up in bed.

Huh?

I hear a voice in my head. *You've got to bring over the second term from the other side, take the Fourier transform, and invert in* L^2.

Unbelievable!

Scribble a note to myself on a scrap of paper, holler at the kids to get dressed, go down to make them breakfast. Hurry up or you'll be late for school. Scampering out the door, skipping over the wet grass and beating me to the bus stop, they take their place in line behind the other children, patiently waiting to board a beautiful yellow-and-black bus—just like the ones in American movies!

Remarkable how many sons and daughters of top scientists take this bus to Littlebrook Elementary School every morning. Look, there are Ngô Bảo Châu's kids! Ngô left Paris several years ago to accept a five-year visiting appointment at the IAS. He's already famous in the mathematical community for his spectacular solution of a long-standing problem called the Fundamental Lemma, now being checked and verified. The branch of mathematics in which Ngô works, algebraic geometry, is renowned for its difficulty—and totally foreign to me. Everyone considers him an odds-on favorite for one of the next Fields Medals!

At Littlebrook the students are pampered. Private English lessons, instruction in other subjects tailored to their individual needs, etc. Great efforts are made to give them a sense of self-confidence—something American teachers can certainly be trusted to do. This afternoon the kids will come back from school in good spirits, actually looking forward to their homework. Lucky thing that the hatred of homework hasn't yet reached the United States, at least not Princeton.

Once the kids are safely on their way I rush back home and settle down in an armchair to try out the idea that came to me when I woke up, as if by magic. Mumbling to myself all the while: "I stay in Fourier, as Sigal recommended, completely avoiding the Laplace transform. But before inverting, I begin by separating thus. Next, in a second step . . ."

I go on scribbling, then pause for a moment to reflect.

"It works! I think. . . ."

"YES! It works!!! *Of course* that's the way you've got to do it, no question. We can elaborate later, add more details. But now the framework's finally in place."

From here on it's only a matter of patience. Developing the idea will undoubtedly lead to schemes that I know what to do with. I write out the details carefully, taking my time. The moment has finally come to bring to bear all the ingenuity I can muster from eighteen years of doing mathematics!

"Hmmmm, now that resembles a Young inequality . . . and then it's like proving Minkowski's inequality . . . you change the variables, separate the integrals. . . ."

I went into semi-automatic pilot, drawing on the whole of my accumulated experience . . . but in order to be able to do this, first you've got to tap into a certain line—the famous direct line, the one that connects you to God, or at least the god of mathematics. Suddenly you hear a voice echoing in your head. It's not the sort of thing that happens every day, I grant you. But it *does* happen.

I'd tapped into the direct line once before. In the winter of

2001, when I was teaching in Lyon, I lectured every Wednesday at the Institut Henri Poincaré in Paris. On one of those Wednesdays I was explaining my quasi-solution to Cercignani's conjecture when Thierry Bodineau interrupted me and asked whether a certain part of the statement couldn't be improved. Thinking about it on the high-speed train back to Lyon, the TGV, divine inspiration or something very much like it showed me the way to a much more powerful proof scheme, which made it possible to give a complete demonstration of the conjecture. Then, over the next few days, I was able to broaden the conjecture by extending the argument to cover a more general case.

But come the following Tuesday, just as I was about to proudly present my new results, I discovered a fatal error in the proof of the second theorem! I worked until three or four in the morning trying to fix it. No success.

Scarcely awake after only a few hours' sleep, I began mulling over the problem again. I couldn't bear the thought of not being able to announce my results. When I left home to go to the station, my head was filled with dead ends. But no sooner had I settled into my seat on the TGV than inspiration suddenly struck again: I *knew* what I had to do to fix the proof.

I spent the rest of the trip from Lyon to Paris putting everything in order, and when I entered the lecture hall I was able to keep my promise to myself after all. This made-in-TGV proof furnished the basis for one of my best articles only a few months later.

Now, once again, on this morning, the morning of April 9, 2009, another bit of inspiration came knocking at my brain's door and illuminated everything!

Probably no one who read the article that finally appeared in *Acta Mathematica* had the least inkling of the euphoria I experienced that morning. Technique is the only thing that matters in a proof. It's a pity there's no place for the most important thing of all: illumination.

•

7.4. **Growth control.** To state the main result of this section we shall write $\mathbb{Z}_*^d = \mathbb{Z}^d \setminus \{0\}$; and if a sequence of functions $\Phi(k,t)(k \in \mathbb{Z}_*^d, t \in \mathbb{R})$ is given, then $\|\Phi(t)\|_\lambda = \sum_k e^{2\pi\lambda|k|}|\Phi(k,t)|$. We shall use $K(s)\Phi(t)$ as a shorthand for $(K(k,s)\Phi(k,t))_{k \in \mathbb{Z}_*^d}$, etc.

Theorem 7.7 (Growth control via integral inequalities). *Let* $f^0 = f^0(v)$ *and* $W = W(x)$ *satisfy condition* **(L)** *from Subsection 2.2 with constants* C_0, λ_0, κ; *in particular* $|\tilde{f}^0(\eta)| \leq C_0 e^{-2\pi\lambda_0|\eta|}$. *Let further*

$$C_W = \max\left\{\sum_{k \in \mathbb{Z}_*^d}|\hat{W}(k)|, \sup_{k \in \mathbb{Z}_*^d}|k||\hat{W}(k)|\right\}.$$

Let $A \geq 0$, $\mu \geq 0$, $\lambda \in (0, \lambda^*]$ *with* $0 < \lambda^* < \lambda_0$. *Let* $(\Phi(k,t))_{k \in \mathbb{Z}_*^d, t \geq 0}$ *be a continuous function of* $t \geq 0$, *valued in* $\mathbb{C}^{\mathbb{Z}_*^d}$, *such that*

$$\forall t \geq 0, \quad \left\|\Phi(t) - \int_0^t K^0(t-\tau)\Phi(\tau)d\tau\right\|_{\lambda t + \mu}$$

$$\leq A + \int_0^t\left[K_0(t,\tau) + K_1(t,\tau) + \frac{c_0}{(1+\tau)^m}\right]\|\Phi(\tau)\|_{\lambda\tau + \mu}\,d\tau, \quad (7.22)$$

where $c_0 \geq 0$, $m > 1$, *and* $K_0(t,\tau)$, $K_1(t,\tau)$ *are nonnegative kernels. Let* $\varphi(t) = \|\Phi(t)\|_{\lambda t + \mu}$. *Then*

(i) Assume $\gamma > 1$ *and* $K_1 = cK^{(\alpha),\gamma}$ *for some* $c > 0$, $\alpha \in (0, \bar{\alpha}(\gamma))$, *where* $K^{(\alpha),\gamma}$ *is defined by*

$$K^{(\alpha),\gamma}(t,\tau) = (1+\tau)d \sup_{k \neq 0, \ell \neq 0} \frac{e^{-\alpha|\ell|}e^{-\alpha\left(\frac{t-\tau}{t}\right)|k-\ell|}e^{-\alpha|k(t-\tau)+\ell\tau|}}{1+|k-\ell|^\gamma},$$

and $\bar{\alpha}(\gamma)$ *appears in Proposition 7.1. Then there are positive constants* C *and* χ, *depending only on* γ, λ^*, λ_0, κ, c_0, C_W, m, *uniform as* $\gamma \to 1$, *such that if*

$$\sup_{t \geq 0} \int_0^t K_0(t, \tau) d\tau \leq \chi \qquad (7.23)$$

and

$$\sup_{t \geq 0} \left(\int_0^t K_0(t, \tau)^2 \, d\tau \right)^{1/2} + \sup_{\tau \geq 0} \int_\tau^\infty K_0(t, \tau) dt \leq 1, \quad (7.24)$$

then for any $\varepsilon \in (0, \alpha)$,

$$\forall t \geq 0, \quad \varphi(t) \leq CA \frac{(1 + c_0^2)}{\sqrt{\varepsilon}} e^{Cc_0} \left(1 + \frac{c}{\alpha \varepsilon} \right)$$
$$\times e^{CT} e^{Cc(1+T^2)} e^{\varepsilon t}, \qquad (7.25)$$

where

$$T = C \max \left\{ \left(\frac{c^2}{\alpha^5 \varepsilon^{2+\gamma}} \right)^{\frac{1}{\gamma-1}}; \ \left(\frac{c}{\alpha^2 \varepsilon^{\gamma+\frac{1}{2}}} \right)^{\frac{1}{\gamma-1}}; \ \left(\frac{c_0^2}{\varepsilon} \right)^{\frac{1}{2m-1}} \right\}. \quad (7.26)$$

(ii) Assume $K_1 = \sum_{1 \leq i \leq N} c_i K^{(\alpha_i), 1}$ *for some* $\alpha_i \in (0, \bar{\alpha}(1))$, *where* $\bar{\alpha}(1)$ *appears in Proposition 7.1; then there is a numeric constant* $\Gamma > 0$ *such that whenever*

$$1 \geq \varepsilon \geq \Gamma \sum_{i=1}^N \frac{c_i}{\alpha_i^3},$$

one has, with the same notation as in (i),

$$\forall t \geq 0, \quad \varphi(t) \leq CA \frac{(1 + c_0^2) e^{Cc_0}}{\sqrt{\varepsilon}} e^{CT} e^{Cc(1+T^2)} e^{\varepsilon t}, \quad (7.27)$$

where

$$c = \sum_{i=1}^{N} c_i, \quad T = C \max \left\{ \frac{1}{\varepsilon^2} \left(\sum_{i=1}^{N} \frac{c_i}{\alpha_i^3} \right); \ \left(\frac{c_0^2}{\varepsilon} \right)^{\frac{1}{2m-1}} \right\}.$$

Proof of Theorem 7.7. We only treat (i), since the reasoning for (ii) is rather similar; and we only establish the conclusion as an *a priori* estimate, skipping the continuity/approximation argument needed to turn it into a rigorous estimate. Then the proof is done in three steps.

Step 1: *Crude pointwise bounds.* From (7.22) we have

$$\varphi(t) = \sum_{k \in \mathbb{Z}_*^d} |\Phi(k,t)| e^{2\pi(\lambda t + \mu)|k|}$$

$$\leq A + \sum_k \int_0^t \left| K^0(k, t-\tau) \right| e^{2\pi(\lambda t + \mu)|k|} |\Phi(t,\tau)| d\tau$$

$$+ \int_0^t \left[K_0(t,\tau) + K_1(t,\tau) + \frac{c_0}{(1+\tau)^m} \right] \varphi(\tau) d\tau$$

$$\leq A + \int_0^t \left[\left(\sup_k \left| K^0(k, t-\tau) \right| e^{2\pi\lambda(t-\tau)|k|} \right) \right.$$

$$\left. + K_1(t,\tau) + K_0(t,\tau) + \frac{c_0}{(1+\tau)^m} \right] \varphi(\tau) d\tau. \quad (7.28)$$

We note that for any $k \in \mathbb{Z}_*^d$ and $t \geq 0$,

$$\left| K^0(k, t-\tau) \right| e^{2\pi\lambda|k|(t-\tau)} \leq 4\pi^2 |\hat{W}(k)| C_0 e^{-2\pi(\lambda_0 - \lambda)|k|t} |k|^2 t$$

$$\leq \frac{CC_0}{\lambda_0 - \lambda} \left(\sup_{k \neq 0} |k| |\hat{W}(k)| \right) \leq \frac{CC_0 C_W}{\lambda_0 - \lambda},$$

where (here as below) C stands for a numeric constant which may change from line to line. Assuming $\int K_0(t, \tau)d\tau \leq 1/2$, we deduce from (7.28)

$$\varphi(t) \leq A + \frac{1}{2}\left(\sup_{0 \leq \tau \leq t} \varphi(\tau)\right)$$

$$+ C\int_0^t \left(\frac{C_0 C_W}{\lambda_0 - \lambda} + c(1+t) + \frac{c_0}{(1+\tau)^m}\right)\varphi(\tau)d\tau,$$

and by Gronwall's lemma

$$\varphi(t) \leq 2Ae^{C\left(\frac{C_0 C_W}{\lambda_0 - \lambda}t + c(t+t^2) + c_0 C_m\right)}, \tag{7.29}$$

where $C_m = \int_0^\infty (1+\tau)^{-m}\,d\tau$.

Step 2: L^2 *bound.* This is the step where the smallness assumption (7.23) will be the most important. For all $k \in \mathbb{Z}_*^d$, $t \geq 0$, we define

$$\Psi_k(t) = e^{-\varepsilon t}\,\Phi(k,t)e^{2\pi(\lambda t + \mu)|k|}, \tag{7.30}$$

$$\mathcal{K}_k^0(t) = e^{-\varepsilon t}\,K^0(k,t)\,e^{2\pi(\lambda t + \mu)|k|}, \tag{7.31}$$

$$R_k(t) = e^{-\varepsilon t}\left(\Phi(k,t) - \int_0^t K^0(k, t-\tau)\Phi(k,\tau)d\tau\right)$$

$$\times e^{2\pi(\lambda t + \mu)|k|}$$

$$= \left(\Psi_k - \Psi_k * \mathcal{K}_k^0\right)(t), \tag{7.32}$$

and we extend all these functions by 0 for negative values of t. Taking the Fourier transform in the time variable yields $\hat{R}_k = (1 - \hat{\mathcal{K}}_k^0)\hat{\Psi}_k$; since condition **(L)** implies $|1 - \hat{\mathcal{K}}_k^0| \geq \kappa$, we deduce $\|\hat{\Psi}_k\|_{L^2} \leq \kappa^{-1}\|\hat{R}_k\|_{L^2}$, i.e.,

$$\|\Psi_k\|_{L^2(dt)} \leq \frac{\|R_k\|_{L^2(dt)}}{\kappa}. \tag{7.33}$$

Plugging (7.33) into (7.32), we deduce

$$\forall k \in \mathbb{Z}_*^d, \quad \|\Psi_k - R_k\|_{L^2(dt)} \leq \frac{\|\mathcal{K}_k^0\|_{L^1(dt)}}{\kappa} \|R_k\|_{L^2(dt)}. \tag{7.34}$$

Then

$$
\begin{aligned}
\left\| \varphi(t) e^{-\varepsilon t} \right\|_{L^2(dt)} &= \left\| \sum_k |\Psi_k| \right\|_{L^2(dt)} \\
&\leq \left\| \sum_k |R_k| \right\|_{L^2(dt)} + \sum_k \left\| R_k - \Psi_k \right\|_{L^2(dt)} \\
&\leq \left\| \sum_k |R_k| \right\|_{L^2(dt)} \left(1 + \frac{1}{\kappa} \sum_{\ell \in \mathbb{Z}_*^d} \|\mathcal{K}_\ell^0\|_{L^1(dt)} \right). \tag{7.35}
\end{aligned}
$$

(Note: We bounded $\|R_\ell\|$ by $\|\sum_k |R_k|\|$, which seems very crude; but the decay of $\mathcal{K}_k^0$ as a function of k will save us.) Next, we note that

$$
\begin{aligned}
\|\mathcal{K}_k^0\|_{L^1(dt)} &\leq 4\pi^2 |\hat{W}(k)| \int_0^\infty C_0 e^{-2\pi(\lambda_0 - \lambda)|k|t} |k|^2 t \, dt \\
&\leq 4\pi^2 |\hat{W}(k)| \frac{C_0}{(\lambda_0 - \lambda)^2},
\end{aligned}
$$

so

$$\sum_k \|\mathcal{K}_k^0\|_{L^1(dt)} \leq 4\pi^2 \left(\sum_k |\hat{W}(k)| \right) \frac{C_0}{(\lambda_0 - \lambda)^2}.$$

Plugging this into (7.35) and using (7.22) again, we obtain

$$\left\| \varphi(t)e^{-\varepsilon t} \right\|_{L^2(dt)} \leq \left(1 + \frac{CC_0 C_W}{\kappa(\lambda_0 - \lambda)^2} \right) \left\| \sum_k |R_k| \right\|_{L^2(dt)}$$

$$\leq \left(1 + \frac{CC_0 C_W}{\kappa(\lambda_0 - \lambda)^2} \right) \left\{ \int_0^\infty e^{-2\varepsilon t} \left[A + \int_0^t \left[K_1 + K_0 \right. \right. \right.$$

$$\left. \left. \left. + \frac{c_0}{(1+\tau)^m} \right] \varphi(\tau) d\tau \right]^2 dt \right\}^{\frac{1}{2}}. \qquad (7.36)$$

We separate this (by Minkowski's inequality) into various contributions which we estimate separately. First, of course,

$$\left(\int_0^\infty e^{-2\varepsilon t} A^2 \, dt \right)^{\frac{1}{2}} = \frac{A}{\sqrt{2\varepsilon}}. \qquad (7.37)$$

Next, for any $T \geq 1$, by Step 1 and $\int_0^t K_1(t, \tau) \, d\tau \leq Cc(1+t)/\alpha$,

$$\left\{ \int_0^T e^{-2\varepsilon t} \left(\int_0^t K_1(t, \tau) \varphi(\tau) d\tau \right)^2 dt \right\}^{\frac{1}{2}}$$

$$\leq \left[\sup_{0 \leq t \leq T} \varphi(t) \right] \left(\int_0^T e^{-2\varepsilon t} \left(\int_0^t K_1(t, \tau) d\tau \right)^2 dt \right)^{\frac{1}{2}}$$

$$\leq C A e^{c \left[\frac{C_0 C_W}{\lambda_0 - \lambda} T + c(T + T^2) \right]} \frac{c}{\alpha} \left(\int_0^\infty e^{-2\varepsilon t} (1+t)^2 dt \right)^{\frac{1}{2}}$$

$$\leq C A \frac{c}{\alpha \varepsilon^{3/2}} e^{c \left[\frac{C_0 C_W}{\lambda_0 - \lambda} T + c(T + T^2) \right]}. \qquad (7.38)$$

Invoking Jensen and Fubini, we also have

$$\left\{ \int_T^\infty e^{-2\varepsilon t} \left(\int_0^t K_1(t,\tau)\varphi(\tau)d\tau \right)^2 dt \right\}^{\frac{1}{2}}$$

$$= \left\{ \int_T^\infty \left(\int_0^t K_1(t,\tau)e^{-\varepsilon(t-\tau)} e^{-\varepsilon\tau} \varphi(\tau)d\tau \right)^2 dt \right\}^{\frac{1}{2}}$$

$$\leq \left\{ \int_T^\infty \left(\int_0^t K_1(t,\tau)e^{-\varepsilon(t-\tau)} d\tau \right) \right.$$

$$\left. \times \left(\int_0^t K_1(t,\tau)e^{-\varepsilon(t-\tau)} e^{-2\varepsilon\tau}\varphi(\tau)^2 d\tau \right) dt \right\}^{\frac{1}{2}}$$

$$\leq \left(\sup_{t \geq T} \int_0^t e^{-\varepsilon t} K_1(t,\tau)e^{\varepsilon\tau} d\tau \right)^{\frac{1}{2}}$$

$$\times \left(\int_T^\infty \int_0^t K_1(t,\tau)e^{-\varepsilon(t-\tau)} e^{-2\varepsilon\tau}\varphi(\tau)^2 d\tau\, dt \right)^{\frac{1}{2}}$$

$$= \left(\sup_{t \geq T} \int_0^t e^{-\varepsilon t} K_1(t,\tau)e^{\varepsilon\tau} d\tau \right)^{\frac{1}{2}}$$

$$\times \left(\int_0^\infty \int_{\max\{\tau;T\}}^{+\infty} K_1(t,\tau)e^{-\varepsilon(t-\tau)} e^{-2\varepsilon\tau} \varphi(\tau)^2 dt\, d\tau \right)^{\frac{1}{2}}$$

$$\leq \left(\sup_{t \geq T} \int_0^t e^{-\varepsilon t} K_1(t,\tau)e^{\varepsilon\tau} d\tau \right)^{\frac{1}{2}}$$

$$\times \left(\sup_{\tau \geq 0} \int_\tau^\infty e^{\varepsilon\tau} K_1(t,\tau)e^{-\varepsilon t} dt \right)^{\frac{1}{2}}$$

$$\times \left(\int_0^\infty e^{-2\varepsilon\tau} \varphi(\tau)^2 d\tau \right)^{\frac{1}{2}}. \tag{7.39}$$

(Basically we copied the proof of Young's inequality.) Similarly,

$$\left\{ \int_0^\infty e^{-2\varepsilon t} \left(\int_0^t K_0(t,\tau)\varphi(\tau)\,d\tau \right)^2 dt \right\}^{\frac{1}{2}}$$

$$\leq \left(\sup_{t\geq 0} \int_0^t e^{-\varepsilon t} K_0(t,\tau) e^{\varepsilon \tau}\,d\tau \right)^{\frac{1}{2}}$$

$$\times \left(\sup_{\tau\geq 0} \int_\tau^\infty e^{\varepsilon \tau} K_0(t,\tau) e^{-\varepsilon t}\,dt \right)^{\frac{1}{2}}$$

$$\times \left(\int_0^\infty e^{-2\varepsilon \tau} \varphi(\tau)^2\,d\tau \right)^{\frac{1}{2}}$$

$$\leq \left(\sup_{t\geq 0} \int_0^t K_0(t,\tau)\,d\tau \right)^{\frac{1}{2}} \left(\sup_{\tau\geq 0} \int_\tau^\infty K_0(t,\tau)\,dt \right)^{\frac{1}{2}}$$

$$\times \left(\int_0^\infty e^{-2\varepsilon \tau} \varphi(\tau)^2\,d\tau \right)^{\frac{1}{2}}. \tag{7.40}$$

The last term is also split, this time according to $\tau \leq T$ or $\tau > T$:

$$\left\{ \int_0^\infty e^{-2\varepsilon t} \left(\int_0^T \frac{c_0\varphi(\tau)}{(1+\tau)^m}\,d\tau \right)^2 dt \right\}^{\frac{1}{2}}$$

$$\leq c_0 \left(\sup_{0\leq \tau\leq T} \varphi(\tau) \right)$$

$$\times \left\{ \int_0^\infty e^{-2\varepsilon t} \left(\int_0^T \frac{d\tau}{(1+\tau)^m} \right)^2 dt \right\}^{\frac{1}{2}}$$

$$\leq c_0 \frac{C\,A}{\sqrt{\varepsilon}} e^{\left[\left(\frac{C_0 C_W}{\lambda_0 - \lambda} \right) T + c(T+T^2) \right]} C_m, \tag{7.41}$$

and

$$\left\{ \int_0^\infty e^{-2\varepsilon t} \left(\int_T^t \frac{c_0\, \varphi(\tau)\, d\tau}{(1+\tau)^m} \right)^2 dt \right\}^{\frac{1}{2}}$$

$$= c_0 \left\{ \int_0^\infty \left(\int_T^t e^{-\varepsilon(t-\tau)} \frac{e^{-\varepsilon\tau}\, \varphi(\tau)}{(1+\tau)^m}\, d\tau \right)^2 dt \right\}^{\frac{1}{2}}$$

$$\leq c_0 \left\{ \int_0^\infty \left(\int_T^t \frac{e^{-2\varepsilon(t-\tau)}}{(1+\tau)^{2m}}\, d\tau \right) \left(\int_T^t e^{-2\varepsilon\tau}\, \varphi(\tau)^2\, d\tau \right) dt \right\}^{\frac{1}{2}}$$

$$\leq c_0 \left(\int_0^\infty e^{-2\varepsilon t}\, \varphi(t)^2\, dt \right)^{\frac{1}{2}} \left(\int_0^\infty \int_T^t \frac{e^{-2\varepsilon(t-\tau)}}{(1+\tau)^{2m}}\, d\tau\, dt \right)^{\frac{1}{2}}$$

$$= c_0 \left(\int_0^\infty e^{-2\varepsilon t}\, \varphi(t)^2\, dt \right)^{\frac{1}{2}}$$

$$\times \left(\int_T^\infty \frac{1}{(1+\tau)^{2m}} \left(\int_\tau^\infty e^{-2\varepsilon(t-\tau)}\, dt \right) d\tau \right)^{\frac{1}{2}}$$

$$= c_0 \left(\int_0^\infty e^{-2\varepsilon t}\, \varphi(t)^2\, dt \right)^{\frac{1}{2}} \left(\int_T^\infty \frac{d\tau}{(1+\tau)^{2m}} \right)^{\frac{1}{2}}$$

$$\times \left(\int_0^\infty e^{-2\varepsilon s}\, ds \right)^{\frac{1}{2}}$$

$$= \frac{C_{2m}^{1/2}\, c_0}{\sqrt{\varepsilon}\, T^{m-1/2}} \left(\int_0^\infty e^{-2\varepsilon t}\, \varphi(t)^2\, dt \right)^{\frac{1}{2}}. \tag{7.42}$$

Gathering estimates (7.37) to (7.42), we deduce from (7.36)

$$\left\|\varphi(t)e^{-\varepsilon t}\right\|_{L^2(dt)} \le \left(1 + \frac{CC_0C_W}{\kappa(\lambda_0-\lambda)^2}\right)\frac{CA}{\sqrt{\varepsilon}}\left[1 + \left(\frac{c}{\alpha\varepsilon} + c_0C_m\right)\right]$$

$$\times e^{C\left[\frac{C_0C_W}{\lambda_0-\lambda}T+c(T+T^2)\right]} + a\left\|\varphi(t)e^{-\varepsilon t}\right\|_{L^2(dt)}, \quad (7.43)$$

where

$$a = \left(1 + \frac{CC_0C_W}{\kappa(\lambda_0-\lambda)^2}\right)\left[\left(\sup_{t\ge T}\int_0^t e^{-\varepsilon t}K_1(t,\tau)e^{\varepsilon\tau}\,d\tau\right)^{\frac{1}{2}}\right.$$

$$\times\left(\sup_{\tau\ge 0}\int_\tau^\infty e^{\varepsilon\tau}K_1(t,\tau)e^{-\varepsilon t}\,dt\right)^{\frac{1}{2}} + \left(\sup_{t\ge 0}\int_0^t K_0(t,\tau)\,d\tau\right)^{\frac{1}{2}}$$

$$\times\left(\sup_{\tau\ge 0}\int_\tau^\infty K_0(t,\tau)\,dt\right)^{\frac{1}{2}} + \left.\frac{C_{2m}^{1/2}c_0}{\sqrt{\varepsilon}\,T^{m-1/2}}\right].$$

Using Propositions 7.1 (case $\gamma > 1$) and 7.5, as well as assumptions (7.23) and (7.24), we see that $a \le 1/2$ for χ small enough and T satisfying (7.26). Then from (7.43) it follows that

$$\left\|\varphi(t)e^{-\varepsilon t}\right\|_{L^2(dt)} \le \left(1 + \frac{CC_0C_W}{\kappa(\lambda_0-\lambda)^2}\right)\frac{CA}{\sqrt{\varepsilon}}$$

$$\times\left[1 + \left(\frac{c}{\alpha\varepsilon} + c_0C_m\right)\right]e^{C\left[\frac{C_0C_W}{\lambda_0-\lambda}T+c(T+T^2)\right]}.$$

Step 3: *Refined pointwise bounds.* Let us use (7.22) a third time, now for $t \ge T$:

$$e^{-\varepsilon t}\varphi(t) \leq A e^{-\varepsilon t}$$

$$+ \int_0^t \left(\sup_k | K^0(k, t-\tau) | e^{2\pi\lambda(t-\tau)|k|} \right) \varphi(\tau) e^{-\varepsilon\tau}\, d\tau$$

$$+ \int_0^t \left[K_0(t,\tau) + \frac{c_0}{(1+\tau)^m} \right] \varphi(\tau) e^{-\varepsilon\tau}\, d\tau$$

$$+ \int_0^t \left(e^{-\varepsilon t} K_1(t,\tau) e^{\varepsilon\tau} \right) \varphi(\tau) e^{-\varepsilon\tau}\, d\tau$$

$$\leq A e^{-\varepsilon t} + \left[\left(\int_0^t \left(\sup_{k\in\mathbb{Z}_*^d} | K^0(k, t-\tau) | e^{2\pi\lambda(t-\tau)|k|} \right)^2 d\tau \right)^{\frac{1}{2}} \right.$$

$$+ \left(\int_0^t K_0(t,\tau)^2\, d\tau \right)^{\frac{1}{2}} + \left(\int_0^\infty \frac{c_0^2}{(1+\tau)^{2m}} d\tau \right)^{\frac{1}{2}}$$

$$+ \left(\int_0^t e^{-2\varepsilon t} K_1(t,\tau)^2\, e^{2\varepsilon\tau}\, d\tau \right)^{\frac{1}{2}} \left] \left(\int_0^\infty \varphi(\tau)^2\, e^{-2\varepsilon\tau}\, d\tau \right)^{\frac{1}{2}}. \quad (7.44)$$

We note that for any $k \in \mathbb{Z}_*^d$,

$$(| K^0(k,t) | e^{2\pi\lambda|k|t})^2 \leq 16\pi^4 | \hat{W}(k) |^2 \left| \tilde{f}^0(kt) \right|^2 | k |^4\, t^2\, e^{4\pi\lambda|k|t}$$

$$\leq C C_0^2 | \hat{W}(k) |^2\, e^{-4\pi(\lambda_0-\lambda)|k|t}\, | k |^4\, t^2$$

$$\leq \frac{C C_0^2}{(\lambda_0-\lambda)^2} | \hat{W}(k) |^2\, e^{-2\pi(\lambda_0-\lambda)|k|t}\, | k |^2$$

$$\leq \frac{C C_0^2}{(\lambda_0-\lambda)^2} C_W^2\, e^{-2\pi(\lambda_0-\lambda)|k|t}$$

$$\leq \frac{C C_0^2}{(\lambda_0-\lambda)^2} C_W^2\, e^{-2\pi(\lambda_0-\lambda)t};$$

so

$$\int_0^t \left(\sup_{k \in \mathbb{Z}_*^d} \left| K^0(k, t-\tau) \right| e^{2\pi\lambda(t-\tau)|k|} \right)^2 d\tau \leq \frac{C C_0^2 C_W^2}{(\lambda_0 - \lambda)^3}.$$

Then the conclusion follows from (7.44), Corollary 7.4, conditions (7.26) and (7.24), and Step 2. □

TWENTY-EIGHT

Princeton
April 14, 2009

Today I officially accepted the directorship of the IHP.

And we're right on track with the theorem. Twice in the last few days I've worked until four in the morning, my resolve undiminished.

This evening I was getting ready for another long private session with the Problem. The first step is always to boil some water.

Then suddenly I realized there wasn't any more tea in the house—panic! Without the stimulating leaves of *Camellia sinensis*, I couldn't possibly face the hours of calculation that lay in store.

Night had already fallen, futile to imagine finding a store open in town. Untroubled by the thought of crime, I hopped on my bike and set off to steal some tea bags from the common room of the School of Mathematics.

Made it to Simonyi Hall in no time flat, typed in the entry code, and furtively climbed the stairs to the second floor. All was darkness, except for a ray of light glimmering beneath Jean Bourgain's door. I wasn't the least bit surprised: even though Jean has won the highest honors and is universally regarded as one of the most powerful analysts of recent decades, he has kept the working hours of an ambitious young up-and-comer—in part, too, because he regularly visits the West Coast and likes to stay on Pacific Standard Time. It's always a good bet you'll find him working late into the night.

Slipped into the common room and noiselessly pocketed the precious packets. André Weil staring down at me in disapproval. I hurried back downstairs.

But there, standing right in my way, was Tom Spencer, a big name in statistical physics and one of my best friends at the Institute. I had no choice but to confess to my burglary.

"Oh, tea! Keeps you going, eh?"

Back home in a flash. The time had come at last to perform my private tea ceremony.

And some music, please—or I shall die!

I've been listening to a lot of singers lately. Catherine Ribeiro, whose songs play in a continuous loop on my computer. The tragic Danielle Messia, the forsaken one. Ribeiro, *la pasionaria*, hopelessly committed to her cause, come what may. The hypersensitive Mama Béa Tekielski, with her magnificent shrieking and screeching. Ribeiro, Ribeiro, Ribeiro. Music, my indispensable companion in solitary research.

Nothing brings back long-forgotten moments in our lives quite the way music does. I remember the shocked look on my grandfather's face the first time he heard me play a piece by Francis Poulenc: instantly he was transported back sixty years in time, back to the modest apartment whose wallpaper-thin walls were powerless to stifle the sounds produced by his next-door neighbor, a classical composer inspired by the same aesthetic ideas as Poulenc.

For me it's no different. When I hear Gundula Janowitz launch into Schubert's "Gretchen am Spinnrade," I become once more the young man hospitalized for pneumothorax in the intensive care unit of the Hôpital Cochin in Paris who spent part of his days devouring *Carmen Cru* and part of his nights talking about music with the interns, sleeping with an Irish teddy bear a girl had given him.

Hearing Tom Waits rasp his way through "Cemetery Polka" takes me back to the time of my second pneumothorax, in a large hospital in Lyon where I shared a room with a guy whose ribald banter made the nurses roar with laughter.

John Lennon's metamorphosis into the Walrus ushers me into a hallway at the École Polytechnique in Paris after the first of my two oral examinations. Eighteen years old—the future drawing an elegant question mark above my head.

Three years later, the dramatic opening chords of Brahms's First Piano Concerto sounded at exactly the right moment in my little dormitory room at the École Normale Supérieure when a girl knocked on my door, distraught, looking for answers.

Going back further, to my early childhood, several pieces of music bring back especially vivid memories: Jeannette's insistent "Porque te vas," the song that made her famous; Steve Waring's gently sarcastic "Baleine bleue"; Henri Tachan's caustic "Grand Méchant Loup"; also (no idea why) a theme from Beethoven's Violin Concerto that my mother liked to hum.

From when I was twelve, some of my parents' favorite songs: Jean Ferrat's "Les Poètes," Maxime Le Forestier's "Éducation sentimentale," Leonard Cohen's "Seems So Long Ago, Nancy," Beau Dommage's "La Plainte du phoque," two songs by Les Enfants Terribles, "Horloge du fond de l'eau" and "Sur un fil blanc," Jean-Michel Jarre's "Oxygène," and Graeme Allwright's "Jusqu'à la ceinture," in which an army captain orders his men to keep going forward even though the water is waist-high and rising.

And from when I was a teenager, choosing among the many music videos I watched on M6 and the many cassettes I picked up here and there along the way, a medley of my own favorite songs: "Airport," "Envole-moi," "Tombé du ciel," "Poulailler's Song," "Le Jerk," "King Kong Five," "Marcia Baïla," "Lætitia," "Barbara," "L'Aigle noir," "L'Oiseau de nuit," "Les Nuits sans soleil," "Madame Rêve," "Sweet Dreams (Are Made of This)," "Les Mots bleus," "The Sounds of Silence," "The Boxer," "Still Loving You," "L'Étrange comédie," "Sans contrefaçon," "Maldòn'," "Changer la vie," "Le Bagad

de Lann-Bihoué," "Aux sombres héros de l'amer," "La Ligne Hol-
worth," "Armstrong," "Mississippi River," "Le Connemara," "Sidi
H'Bibi," "Sunday Bloody Sunday," "Wind of Change," "Les Murs de
poussière," "Mon Copain Bismarck," "Hexagone," "Le France,"
"Russians," "J'ai vu," "Oncle Archibald," "Sentimental Bourreau" . . .

•

So many times I've been utterly captivated by a piece of music, classical or pop or rock; so many times I've listened to a piece of music
over and over again, in some cases hundreds of times, marveling at
the state of grace that must have presided over its creation. My entry
into the new world of so-called classical music began with Dvořák's
New World Symphony. After that came Bach's Fifth Brandenburg
Concerto, Beethoven's Seventh Symphony, Rachmaninov's Third
Concerto, Mahler's Second Symphony, Brahms's Fourth Symphony,
Prokofiev's Sixth Sonata, Berg's First Sonata . . . Liszt's Sonata,
Ligeti's Études for Piano, Shostakovich's ambiguous Fifth Sym-
phony, Schubert's Sonata D.784, Chopin's Sixth Prelude (with a suit-
ably dramatic interpretation, thank you very much). Boëllmann's
Toccata, Britten's *War Requiem*, John Adams's fabulous *Nixon in
China*, the Beatles' "A Day in the Life," the Zombies' "Butcher's
Tale," the Beach Boys' "Here Today," Divine Comedy's "Three Sis-
ters," the Têtes Raides' "Gino," Anne Sylvestre's "Lisa la Goélette,"
William Sheller's "Excalibur," Thomas Fersen's "Monsieur."
Étienne Roda-Gil's faux-lighthearted "Ce n'est rien," his faux-faux-
serious "Makhnovchina," his "Le Maître du Palais" (with its palace
of tartaric columns), and his "Patineur" (to the north, or possibly
the south, of July). François Hadji-Lazaro singing of dikes, barges,
and Paris at the barricades. Mort Shuman's love letter to Brooklyn
by the Sea and Herbert Pagani's ode to Venice, now in danger of being drowned by a different sea. Léo Ferré's mysterious, reorches-
trated version of "Inconnue de Londres," and the rabid dog of his

"Chien," the only one that will be left when "Il n'y a plus rien." Bob Dylan in his watchtower recounting the terrible fate of John Brown; Pink Floyd waxing nostalgic over the green grass of yesteryear; Ástor Piazzola evoking Buenos Aires at zero hour. Two film soundtracks, Prokofiev's "Romance" and Ennio Morricone's "Romanzo." Salvatore Adamo's moving "Manuel," whose words I once transcribed for friends in Moscow, lovers of French music and the French language, back in the days when song lyrics couldn't be found on the Internet. Fabrizio De André weeping for Geordie, hung by a golden rope; Giorgio Gaber, taking himself for God; Paolo Conte, inviting his sweetheart to follow him. Little René Simard, making mothers in Quebec cry with his crystalline "Oiseau," as well as young girls in Japan with his breathtaking "Non ne pleure pas" / "Midori iro no yane." Les Frères Jacques buying a five-star general in Francis Blanche's song "Général à vendre." Weepers Circus offering love to vixens, Olivia Ruiz repairing broken hearts and windows, Mes Aïeux's moronic crooks adulterating marijuana in "Ton père est un croche." Boris Vian waxing lyrical about a nuclear thrashing, Gilbert Bécaud about a diabolical auction. Renaud reciting the saga of Gérard Lambert, François Corbier the tale of the unfortunate elephant lover. Hubert-Félix Thiéfaine's strange world, populated by "weed-reaper" girls dispensing joints, by coffins on wheels and atomic Alligators and mucous Diogenes that got all the girls and boys dancing at the all-night parties I used to go to when I was in my twenties. Moments of riveting drama: Jacques Brel crying out, nailed to the Big Dipper; Serge Utgé-Royo reviving Jacques Debronckart's banned song "Mutins de 1917"; Jean Ferrat hailing two children who fell in battle, to their mother Maria's undying sorrow; Henri Tachan howling that he doesn't want to have a child! And the elves who take you by surprise: Kate Bush and her "Army Dreamer," France Gall and her "Petit Soldat," Loreena McKennitt and her "Highwayman," Tori Amos imagining herself as the "Happy Phantom," Jeanne Cherhal shouting "Un trait danger," Amélie Morin playfully blaspheming "Rien ne va plus." And my favorites, the tigresses who give you goose bumps: Melanie reproaching all the ones around her, Danielle Messia demanding to know why she has been abandoned, Patti Smith taking refuge in the night that belongs to lovers, Ute Lemper bemoaning the fate of Marie Saunders, Francesca Solleville bringing the Commune back to life, Juliette playing the little boy manqué, Nina Hagen growling Kurt Weill, Gribouille roaring her Ravens' song, the sublime duo of Patrice Moullet and Catherine Ribeiro singing of peace, death, and the bird in front of the door!

No path can be left unexplored when it comes to tracking down new music. Concert listings, online discussion forums, free music sites—and, of course, the wonderful Internet radio station Bide&Musique, which has introduced me to Évariste, Adonis, Marie, Amélie Morin, Bernard Brabant, and Bernard Icher, as well as Stars de la Pub's fantasy of airplane runways on the Champs-Élysées and Dschinghis Khan's disco hymn to the glories of Moscow.

Doing mathematics is no different, really. You're constantly exploring, your eyes and ears are always open, and then every once in a while you're completely smitten by something and you pour your heart and soul into it, you tell yourself over and over again, hundreds and hundreds of times, that nothing else matters. Well, almost nothing else.

Sometimes the two worlds communicate with each other. Certain pieces of music that have kept me going in the course of a project are forever associated with moments of intense emotion.

When I hear Juliette belting out "Monsieur Vénus," I see myself again sitting under a skylight in Lyon in the winter of 2006, writing up my contribution to the *Proceedings* of the International Congress of Mathematicians in Madrid, where I was an invited speaker.

"Comme avant," by the mischievous Amélie Morin, and "Hung Up on a Dream," by the melodious Zombies, take me back to the summer of 2007 and an apartment in Australia, where I went to learn the regularity theory of optimal transport from the leading experts on the subject—and where I became a fan of the adventures of L, M, and N in *Death Note*, but that's another story. . . .

When Marie Laforêt launches into "Pourquoi ces nuages," with the incomparable nuances of a voice that is frail and powerful at the same time, I find myself once again in England, at Reading in the winter of 2003, fathoming the mysteries of hypercoercivity.

An untitled song by the fiery Jeanne Cherhal plunges me back into the Probability Summer School at Saint Flour, a small town in the Massif Central in France, where in 2005 I won the Ping-Pong tournament before a cheering crowd.

I listened to Prokofiev's Second Concerto, whose fourth movement moves me to tears, practically every day during the fall of 1999 in Atlanta while working on my first book on optimal transport.

Mozart's Requiem! I woke up with it every morning when I was taking the *agrégation* exam in 1994.

And Pär Lindh Project's "Baroque Impressions," resounding now and ever more from the depths of a winter's night following a triumphant plenary lecture at a conference in Reykjavik in January 2005. . . .

The hope of discovery and the frustration of imperfection. The proof that remains tantalizingly out of reach. Happiness mixed with pain. The pleasure of feeling alive that songs overflowing with passion go together with so well.

This evening I wasn't somewhere else. I was alive and well in Princeton, and Ribeiro had to be the one by my side as I worked. Impossible to find her in record stores. Fortunately, there's the Web: a few songs on her site, also the extraordinary *Long Box* album available via musicMe.

The astounding "Poème non épique" is beyond anything one

can imagine—an altogether singular achievement in the history of French chanson. But it's *too* emotionally charged. My hair stands on end if I so much as think about it, I couldn't work and listen at the same time.

Instead I chose the magnificent "Jour de Fête." Power, restraint, emotion, evocative force.

Much rather have been somewhere else
But somewhere else didn't exist

Then comes my favorite moment, when the voice, calm and controlled until now, suddenly takes flight and begins to make its full power felt—this voice which, as Ribeiro herself says in another song, makes "the dead, the living dead, and the living tremble."

Didn't feel like eating or drinking
I wanted to make love
Anywhere anyhow
As long as it's really love
Even if it's on the floor
Long as the feeling gets through

Work, Cédric, work. Tea, equations, Ribeiro.

So many sick people that night
Straining so very hard to make it
Sheets tinted a macabre dawn
Smelling, reeking of alcohol

Wow . . .

Once the song was finished I played it again, and again, and again. I needed this loop, needed to go round and round, in order to leap forward. Work, Cédric, work.

•

HOLIDAY

The big day had finally come all right
Festive cheer echoing throughout the night
Behind every window strings of light
Candles gleaming hot like diamonds
'Twas the night before the famous day
We've got to celebrate by making
Long, long lines at the cash register
Oh what a tremendous waste that day—

How brightly Paris sparkled that night
But not for me, I wasn't even there
Me, I'd crossed paths with a satellite
At just the wrong place on my orbit
What the hell was I doing shopping
Stores decked out in all their fake holly
Looking to find some pseudo-rare object
Looking to find some last little surprise—

Much rather have been somewhere else
But somewhere else didn't exist
Didn't feel like eating or drinking
I wanted to make love
Anywhere anyhow
As long as it's really love
Even if it's on the floor
Long as the feeling gets through—

No calls to answer that night
Phone company's fault no doubt

Champagne had lost all its taste
Struggled hard to stay awake
Time showed the way to heartbreak
Rain beating on the windowpanes
Nothing could be more pathetic
Hot body in an empty bed—

So many sick people that night
Straining so very hard to make it
Sheets tinted a macabre dawn
Smelling, reeking of alcohol
The Big Day had come—Day of Peace
Rock bottom of my Americas
I was dreaming of my satellite
At just the wrong place on my orbit—

TWENTY-NINE

Princeton
April 20, 2009

Teacup in hand, the elderly man turns around and stares at me, not saying a word, visibly taken aback by my rather unusual style of dress. . . .

I'm used to people being confused or intimidated by my spider and
the clothes I wear. Usually I regard them with benign amusement. But
in this case I was at least as intimidated as the person looking at me.
For this person was none other than John Nash—perhaps the greatest
analyst of the century, my mathematical hero. Born in 1928, he never
won the Fields Medal, a failure he bitterly brooded over for decades.
Later, of course, he did receive the Nobel Prize for his early work on
what soon became known as Nash equilibria, which brought him
fame in the fields of game theory, economics, and biology. But what
came after this, in the eyes of connoisseurs, was far more extraordinary. It deserved one, or two, maybe even three Fields Medals.

In 1954, Nash introduced the concept of nonsmooth embed-
dings: monstrosities that made it possible to do impossible things,
such as crumpling up a Ping-Pong ball without deforming it, or
constructing a perfectly flat ring. In the words of Mikhail Gromov:
It couldn't be true and yet it was true—this from a man who understands
Nash's geometrical work better than anyone on earth, and who used
it to develop the whole theory of convex integration.

In 1956, proving himself equal to a challenge contemptuously issued by Warren Ambrose, Nash amazed everyone by demonstrating that all the abstract geometries described by Bernhard Riemann—a prince among princes, the Chopin of mathematics—can actually be included ("embedded") in Euclidean space, the space of our own physical world. In so doing he realized a dream that went back more than a century.

In 1958, in response to a question posed by Louis Nirenberg, Nash demonstrated the regularity of solutions of parabolic linear equations with measurable elliptic coefficients (in physical terms, the continuity in space-time of heat in a completely heterogeneous solid). This result marked the beginning of the modern theory of partial differential equations.

It was fate's wish that the monastic genius Ennio De Giorgi should solve this last problem at the same time, independently of Nash and by a completely different method. Evidently this takes nothing away from Nash's remarkable achievement.

Nash may be the only living scientist to have been cast as the hero of a Hollywood film. I didn't much care for the film myself, but I do have a high regard for the biography by Sylvia Nasar on which it was based. John Nash—a beautiful mind indeed!

If Nash attracted Hollywood's attention, it wasn't only on account of his mathematical exploits. It was also because of the tragic story of his life. At the age of thirty he succumbed to paranoid schizophrenia. In and out of psychiatric clinics and hospitals for more than ten years, he seemed fated to live out his days as a pitiable phantom haunting the halls of Princeton, his mind an incoherent ruin.

But then, after three decades of purgatory, Nash miraculously came back from the far shores of madness. Today, more than eighty years old, he is as normal as you or I.

Except that there is an aura about him that neither you nor I have, an aura due to phenomenal accomplishments, strokes of pure genius—and a way of dissecting and scrutinizing problems that makes Nash a model for all modern analysts, myself most humbly among them.

The man who fixed his gaze on me that day is more than a man, he is a living legend. I didn't have the courage to go talk to him.

But the next time our paths cross I shall dare to approach John Nash. And I shall tell him of my talk on the Scheffer–Shnirelman paradox, which emerged from a proof inspired by his own non-smooth embedding theorem. I shall tell him of my intention to give a talk on his work at the Bibliothèque Nationale de France. Perhaps I will even tell him that he is my hero. Would he find that ridiculous?

•

In the fall of 1956, in New York, a tall, strapping young man pushed open the door to an office in an old factory building just off Washington Square. Next to the door, a sign: INSTITUTE FOR MATHEMATICS AND MECHANICS. *The visitor's striking good looks would have given him little reason to envy Russell Crowe, the movie star who was to play him a half century later. His name was Nash, and at the age of twenty-eight he was already famous for his invention of the Nash equilibrium concept and his proof of the embedding theorem, the result of work done first at Princeton and then at MIT. In New York he was to make new acquaintances and discover new problems.*

The question put to him by Nirenberg captured his attention at once. It had so far defied some of the best mathematical minds—a worthy adversary! The continuity of solutions to parabolic equations with discontinuous coefficients.

In 1811 the great Fourier had worked out the heat equation, which governs the evolution of temperature as a function of position and time in a homogeneous solid as it cools:

$$\frac{\partial T}{\partial t} = C \, \Delta T.$$

Fourier's equation subsequently became one of the most distinguished representatives of the class of partial differential equations. These equations describe all the continuous phenomena of our physical world, from ocean currents to quantum mechanics.

Even if a homogeneous solid is heated in a very uneven manner, so that at any given instant the temperature varies abruptly and erratically from one place to another, the solid has only to be allowed to cool for a fraction of a second for the temperature distribution to smooth out, that is, to vary in a regular manner. This phenomenon, known as parabolic regularization, is one of the first things students learn about in an introductory course on partial differential equations. The corresponding mathematical statement enjoys an importance that goes well beyond the field of physics.

If the solid is inhomogeneous, however, being composed of various materials, at each position x it will have a more or less great conductivity C(x), which is to say a more or less great capacity for cooling. As a result, the equation changes:

$$\frac{\partial T}{\partial t} = \nabla \cdot \Big(C(x) \nabla T \Big).$$

Does the property of regularization still hold true in this context?

Nash, unlike Nirenberg, had no special expertise in PDEs. Just the same, he took the bait. Week after week he kept coming back to discuss the problem with Nirenberg—and to pick his brains.

At first Nash's questions were naïve, questions a beginner might ask. Nirenberg started to wonder whether Nash really deserved his reputation. It takes courage, when you are already famous and admired, to act the novice in a field you haven't yet mastered—courage, or else a quite unusual degree of self-confidence! The willingness to risk the embarrassment of appearing less than brilliant in the hope that the answer to an unexpected query will point you in the right direction. But this is the price of making progress on a difficult problem. . . . Little by little Nash's questions became more precise, more pertinent. An idea was beginning to form in his mind.

Later, back at MIT, Nash continued to work on the problem. Always the same modus operandi: prying information out of one colleague, asking advice from another, explaining the obstacles he had encountered to someone else.

Lennart Carleson, a very talented Swedish analyst, spoke to him about Boltzmann and entropy. Carleson was one of the few mathematicians at the time who was well versed in the subject—not quite by accident, as it happens, since he was Torsten Carleman's literary executor. Carleman was the first mathematician to really tackle Boltzmann's equation in earnest. At the time of his death he had left an unfinished manuscript on the topic, and it was Carleson's job to complete it and see it through publication. This is how Carleson became familiar with the notion of entropy, and why he was able to help Nash.

But Boltzmann and Fourier aren't at all the same thing—entropy and regularity have nothing to do with each other!

Somehow a light went on in Nash's brain. A plan of attack began to take shape. Without showing his cards, the young mathematician continued to make his rounds, knocking on doors, picking up a lemma here, a proposition there.

John Nash

Finally, one morning, the reality of the situation was obvious to all: by combining all the contributions of his various collaborators with unsurpassable skill, Nash had proved the theorem—like an orchestra conductor, or a film director, who gets each player to play his part of the score, or the script, just perfectly.

At the heart of the proof was entropy. Under Nash's direction, entropy was cast against type to tremendous effect. His manner of using differential inequalities involving certain quantities, inspired by an interpretation that was half mathematical, half physical, founded a new style of analysis, and with it a tradition that I am proud to carry on in my own way.

THIRTY

My neck touches the carpet—and instantly a wave of well-being radiates throughout my body, from my head to my toes. Back to my office after lunch, 1:00 p.m., maybe a bit later: just the right moment for a bit of relaxation.

Not the violent kind of relaxation that the astrophysicists in the building next door are so excited about, but not a gentle one either. The only soft thing between me and the floor of my modest office is a thin-pile carpet. Still, you can feel the carpet against your neck, and once you get used to it you wouldn't want it to be any thicker than it is.

Images pass before my closed eyes, one after the other, sounds chirping in my ears, louder and louder. The whole morning replays itself in my head. . . .

A class from the Littlebrook Elementary School came to visit the Institute today. The pond, the magnificent trees in blossom, the great bust of Albert Einstein in the old library—look, kids, the magic castle of science! Eight years old isn't too young to be dreaming about great scientists.

I'd prepared a twenty-minute talk. I told them about Brownian motion, which reveals the world of atoms; and then explained the famous Syracuse problem, which is so simple that a child of eight can understand it—and so complex that the best mathematician in the world throws up his hands in despair.

They listened attentively, staring in wonder at the marvelous images of Brownian movement roaming across the screen of my laptop, which I held up for all to see. In the last row, a wide-eyed fair-haired little boy listened still more attentively than the others. He had been living here only four months, and yet he had no trouble understanding what his father was saying in English with a French accent so thick you could cut it with a knife.

Then the rest of the morning, then another good lunch, then— and then my brain began to fog over, as it often does this time of day. Time to make a fresh start, a clean break—a reboot, as I call it. You restart the computer, clearing its memory and beginning all over again. . . .

Ears ringing, the children chattering away, everything spinning around. Facial muscles relax; the ringing more intense now, sentence fragments flapping about, some more loudly than others; voices and songs, the meal once more, a forgotten spoon, a formal welcome, an unfrozen lake, a bust in my library, $3n+1$, $3n+2$, $3n+3$, the parquet and the shadows and you forgot a small child and . . .

A sudden slight tremor in my limbs, the mists disperse. My mind is clear once more.

I remain still for another moment, stretched out on the floor. Some ants are scurrying across the soles of my feet. . . .

My feet have disappeared from my internal radar. Incredibly heavy, impossible to move. Like when a clump of snow gets stuck to the bottom of your ski and you can't lift it up.

And yet on my first try, as if by magic, movement is restored to my feet, and once again I am whole. The break is over, ten minutes exactly. I'm a new mathematician.

Cedric reboot (completed)

A new Cédric. Time to embark on a fresh round of calculations. There's also an article on Landau damping, more than fifty years old but still very relevant today, that I just got from the library. Now for two hours of intense concentration before tea . . .

•

The Syracuse problem (also known as the Collatz conjecture or the $3n + 1$ prob-lem) is one of the most famous unsolved enigmas of all time. Paul Erdős himself is on record as saying that mathematics is not yet ready to confront such monsters.

Enter the expression "$3n + 1$" in an Internet search engine and follow the thread back to the abominable problem and its result, as simple and insistent as the refrain of a pop song:

Take any whole number you like, say 38.

This number is even. Divide it by 2 *and you get* 19.

This last number is odd. Multiplying by 3 *and adding* 1 *you get* $19 \times 3 + 1 = 58$.

This last number is even. Divide it by 2. . . .

And so on. You go on from one number to the next by means of a simple rule: each time you encounter an even number, divide by 2; *each time you encounter an odd number, multiply by* 3 *and add* 1.

Starting from 38, *as in the example above, you get the following sequence:* 19, 58, 29, 88, 44, 22, 11, 34, 17, 52, 26, 13, 40, 20, 10, 5, 16, 8, 4, 2, 1, 4, 2, 1, 4, 2, 1, 4, 2, 1, 4, 2, 1, 4, 2, 1 . . .

Once you arrive at 1, *in other words, you know what comes next:* 4, 2, 1, 4, 2, 1, 4, 2, 1 . . . , ad calculam aeternam.

Every single time this calculation has been performed in the course of human history, it has ended up at 4, 2, 1. . . . *Does that mean it will* always *turn out thus, no matter which number is chosen as the point of departure?*

Since there are infinitely many integers, obviously it's impossible to try all of them. With all the pocket calculators, desktop calculators, computers, and supercomputers at our disposal today, it has nevertheless been possible to try billions and billions of them, and every last one has wound up leading back to the implacably repeating 4, 2, 1 pattern.

Mathematics is democratic, of course, and anyone is free to try to show that this sequence embodies a general rule. Everyone believes the rule to be true, but since no one knows how to prove it, it remains a conjecture. Whoever succeeds in confirming or disconfirming this conjecture will be proclaimed a hero.

I am certainly not among those who will try. Apart from the fact that it seems to be phenomenally difficult, it isn't the sort of problem that suits the way my mind works. My brain isn't used to thinking about such things.

•

Date: Mon, 4 May 2009 17:25:09 +0200

From: Cedric Villani <Cedric.VILLANI@umpa.ens-lyon.fr>

To: Clement Mouhot <clement.mouhot@ceremade.dauphine.fr>

Subject: Backus

So here's Backus's article from JMP 1960 (Vol.1 No.3, too bad it wasn't Vol.1 No.1!)

Fantastic! Take a look at the next-to-last section of Backus's article, and then the last sentence of the article. It's all the more remarkable since as far as I'm aware no one has explicitly expressed these doubts until recently, the last few years . . .

Best

Cedric

Date: Sun, 10 May 2009 05:21:28 +0800

From: Clement Mouhot <clement.mouhot@ceremade.dauphine.fr>

To: Cedric Villani <Cedric.VILLANI@umpa.ens-lyon.fr>
Subject: Re: Backus

I read a bit of Backus's article on the plane. It is indeed
very interesting, he had a good grasp of the linear pbs and
the question of the increase over time of the background
term once it depends on the space, through filamentation.
And for the most part he's remarkably rigorous compared to
the "standard" of articles on Landau damping . . . We need
to cite to him, particularly his numerical discussion page
190, and the conclusion expressing his doubts regarding the
nonlinear validity of the linear case: this links up with
one of the conceptual difficulties of our intro as well.

Best regards, Clement

THIRTY-ONE

Princeton
A lovely evening in May 2009

May at the Institute for Advanced Study. The trees are in bloom, it's magnificent.

Night has just fallen. I'm wandering alone in the half-light, savoring the encroaching darkness, the sense of peace, the softness of the air.

When I was a student at the École Normale Supérieure, I used to love to walk the dark corridors of the dormitory at night, a few rays of light glimmering beneath the doors, like the vague luminescence one imagines passing through the portholes of the submarine in Jules Verne's famous tale.

But there's no comparison with the evening here, the lawns and the breeze. You've got the light, too, but it isn't the artificial illumination of civilized life, it's the natural light emitted by fireflies— uncountably many twinkling stars covering the grass!

Oh, wait a minute . . . didn't I read an article once, applying Landau's damping theory to the twinkling of fireflies? . . .

*Cédric, please! Can't you forget about Landau damping for just a little while? Enough already—*mille pompons! *You've spent so many days and nights on it as it is. Come on, enjoy the fireflies and give Landau a rest—*

Who's that coming toward me? Apparently I'm not the only one out for a walk after sunset . . . I recognize the silhouette . . . well, what do you know! It's Vladimir—Vladimir Voevodsky, one of the most brilliant mathematicians of his generation, 2002 Fields medalist, one of Grothendieck's spiritual heirs. Just the sort of person you're apt to run into if you venture out at night at the IAS.

Like me, Voevodsky was out walking with no particular destination in mind, just walking, walking for air—like the pedestrian in Ray Bradbury's story.

We got to talking. It's hard to imagine someone whose mathematical interests are more different from mine. I don't understand a thing about his research, and he probably doesn't understand a thing about mine. Rather than try to explain to me what he's doing now, Vladimir told me about what he wants to do next. He's completely fascinated by expert languages and automated theorem proving, and plans to devote himself wholeheartedly to them in the years ahead.

Vladimir Voevodsky

He talked about the famous four-color theorem, and the disputed proof by Appel and Haken—disputed because, in the eyes of their critics, their proof had "dehumanized" mathematics by subjugating it to computer science. The controversy was recently dispelled (or aggravated, depending on your point of view) by French researchers who brought about a small revolution with the aid of an expert language known as Coq.

Vladimir thinks the time isn't far off when computer programs will be able to check long and complex arguments. In fact, he says, such programs are now being tried out in France on a number of famous results. I must admit I was skeptical at first. But the person telling me this isn't mad; he's quite sane, and a mathematician of the highest ability. I've got to take what he says seriously.

Proof checking is the sort of problem I've never even looked at. Algorithmics in general is something I know very little about. It is true, of course, that so-called marriage algorithms (technically, bipartite matching algorithms), simplex algorithms, and auction algorithms play an important role in the numerical simulation of optimal transport, one of the topics that I specialize in; but in optimal transport they're used in a way that is very different from what Vladimir's talking about. This new field really does seem exciting. There are so many wonderful things to explore today!

Flowers, languages, four colors, marriage—everything you need to write a fine song . . . unless perhaps someone's already written it?

•

Around 1850 the mathematician Francis Guthrie colored a map of the counties of England, taking great care to ensure that no two counties sharing a boundary of any length have the same color. How many colored pencils would he have had to use?

Guthrie saw that four colors sufficed. What is more, he suspected that four would suffice to color any such political map (any map, that is, that does not contain counties or states or countries that are divided into noncontiguous parts).

Three colors are plainly not enough. Look at the map of South America, for example, and you will see that in the case of Brazil, Argentina, Bolivia, and Paraguay, each of these four countries borders on three others, and so you need at least four different colors.

The conjecture that four are all you will ever need is something that you can test yourself by coloring your favorite map. If you had enough patience, you could test it for a rather large number of examples. But how could it be shown to be true for every *map? Nobody can possibly test every one, since once again there are an infinite number. There is no alternative but to construct an argument on purely logical grounds. As it turns out, this is not easily done.*

In 1879, Alfred Kempe believed he had proved the four-color theorem. But his proof was flawed, and demonstrated only that five colors suffice.

Let's take it one step at a time. For a map with four countries, we know what the answer is. Beyond that, it's a simple matter to find the answer for five countries, likewise for six. How far can we go on before we run into difficulties?

Suppose that we know all maps up to 1,000 countries can be colored using four colors and that we want to try our luck with a map representing 1,001 countries. How should we proceed? To begin with, it can be shown that among these 1,001 countries there exists at least one that has relatively few immediate neighbors, let's say five at most. If we restrict our attention to this country and the ones adjacent to it, there's no problem; and if we play the conqueror and do a bit of rearranging within the group as a whole, a merger here, a recombination there, we will end up with a map having fewer than 1,000 countries—and so we will be able to color it with four colors. A clever idea . . . but getting the local coloring and the global coloring to coincide is a complicated business. One still has to consider a great many cases—millions, indeed billions of cases!

In 1976, Kenneth Appel and Wolfgang Haken managed to reduce the number of configurations needing to be tested to a bit less than 2,000, and went through all of them with the aid of a computer program. After running the program for two months they concluded that four colors will always suffice, thus resolving a conjecture first stated more than a hundred years earlier.

The mathematical community was profoundly divided over this proof. Hadn't the mind finally been vanquished by the machine? Could any human intelligence really understand an argument that had served as fodder for a monstrous creature made of silicon and integrated circuits? Supporters and opponents squared off, but no consensus emerged.

Eventually mathematicians and computer scientists got used to living with this unsettled state of affairs. Fast forward now to the turn of the millennium and the research being conducted by an INRIA team led by Georges Gonthier, a leading authority on proof-checking languages. The fields of computer science and computational science had been pioneered in Europe by a few visionary theorists at the same time that Appel and Haken were making headlines. The languages they devised were meant to check the soundness of a mathematical proof in the same way you would check the health of a tree: branch by branch. Imagine an argument having the form of a logic tree whose branchings can be subjected to automated verification, just as the correctness of your spelling can be decided by a spell-check program.

But whereas a spell checker is interested only in making sure that individual words are properly formed, a proof checker is designed to check the consistency of the proof as a whole, to make sure that everything hangs together—that each statement follows from the one before without contradiction.

Gonthier and his principal collaborator, Benjamin Werner, set to work trying to prove the four-color theorem using a language called Coq (after the name of its inventor, Thierry Coquand). Unlike the programs used by Appel and Haken, Coq is "certified"—known, that is, to be bug-free. Another difference is that Coq doesn't provide you with the actual computations, it automatically generates a proof on the basis of the algorithm that has been selected. Gonthier took advantage of this fact to rewrite the "readable" part of the proof, and in this way succeeded in obtaining a simple and effective result—a thing of beauty! What you have, then, is a proof of which 0.2% was done by a human being and the other 99.8% by a machine. But it is the human 0.2% that matters, that must be gotten right, since with Coq you can be sure that the rest is correct.

The work of Gonthier and his team heralds the day, not as far in the future as one might suppose, when validation programs will control rocket launchings, airplanes in flight, even the microprocessors in our personal computers. The financial stakes of what was no more than wishful thinking thirty years ago are now reckoned in the billions of dollars.

In the meantime, the indefatigable Gonthier has embarked on an extremely ambitious project aimed at verifying certain classification theorems of finite groups, whose proofs, among the longest of the twentieth century, are renowned for their complexity.

•

A word that goes against
A word that looks askance
A flame shall grow immense
Throughout the universe
Mouths opened wide as pyres
Fine phrases sweet as rum
Yet hides find their buyers
Over the head of a drum

One day will our language
Speak truly of flowers
And also of the marriage
Of four brilliant colors
Will it ever come to you
How they speak of love's flower
I promise I shall wait for you
I shall wait beneath the Tower

Cain shall meanwhile go on chasing Abel
But with my bare hands I built the Tower of Babel

[Guy Béart, from "La Tour de Babel"]

THIRTY-TWO

My last day in Princeton. Rain, nothing but rain the last few weeks—so much, in fact, it was almost comical. At night the fireflies transform the oaks and red maples into Christmas trees, romantic, impossibly tall, decorated with innumerable blinking lights. Enormous mushrooms, a small furtive rabbit, the fleeting silhouette of a fox in the night, the startling noises of stray rutting deer . . .

In the meantime a lot has happened on the Landau damping front! We were finally able to get the proof to hang together, went through the whole thing one last time to be sure. What a wonderful feeling, finally to post our article online! As it turned out, it was in fact possible to control the zero mode. And Clément discovered that we could completely do without the double time-shift, the trick I came up with on my return from the Museum of Natural History back in January. But since we didn't have the courage to go over the whole thing *yet again*, and since we figured it might come in handy dealing with other problems, we left it where it was. It's not in the way, not interfering with anything. We can always simplify later if we have to.

By this point I've given quite a few talks about our work. Each time I was able to improve both the results and the exposition, so the proof is now in very good shape. There may still be a bug somewhere, of course. For the moment, however, all the pieces fit together so well that if an error is discovered, I'm confident we'll be able to fix it.

I was invited to speak for two hours at the Plasma Physics Laboratory in Princeton, and afterward I was treated to a marvelous tour of the lab's facilities. A chance to see the equipment being used to decipher the mysteries of plasmas—and perhaps one day to tame nuclear fusion, who knows?

At the University of Minnesota, in Minneapolis, my presentation seemed to impress Vladimir Šverák. I have the greatest respect for this man, who understands the mysterious notion of quasi-convexity better than anyone and who is now one of the leading authorities on Navier–Stokes regularity. His warm words were a great source of encouragement to me.

Following my talk, at the colloquium dinner, I won over an even tougher audience: the very young, very blond, and very shy daughter of another fine mathematician teaching at Minnesota, Markus Keel. She was delighted to show off her gymnastic skills, executing forward rolls and roaring with laughter. Markus couldn't get over the fact that his daughter was willing to play with a total stranger—she never says a word to people she doesn't know!

At Rutgers I presented my results once more, this time at the spring Statistical Mechanics Conference, one of two organized every year by the indefatigable Joel. It wasn't anything like my previous talk the end of January. This time the logic was tight.

Finally, at the IAS, I gave a lecture to an audience almost exclusively made up of young women as part of the annual Program for Women and Mathematics. They had come here in the hope of lifting the curse that has made mathematics an overwhelmingly male discipline—less so than computer science or electrical engineering, but mostly male just the same. With luck and hard work, some of them may turn out to be worthy successors of the great female pioneers who have been the wonder of mathematics for generations, the next Sofia Kovalevskayas, Emmy Noethers, Olga Oleiniks, Olga Ladyzhenskayas. The young women taking part in the program this year have been a breath of fresh air. In the evenings they can be seen walking about the grounds of the Institute in small groups, enjoying the cool evening breezes.

Yesterday I went with Claire and the children to say our good-byes to the golf course. I have fond memories of walking through the course on my way home at night along the path that leads from the little train station on the edge of the Princeton campus to the Institute, all alone, under a lustrous moon that transformed the sand traps into ghostly waves. . . . And the kids have their own memories: with great reverence they deposited their precious offering on the grass—all the lost golf balls they'd recovered since they got here six months ago! Think of it, six months gone by already!

The state of mathematical grace in which I had been living almost from the beginning of my stay in Princeton lasted until the very end. Once the Landau damping problem was solved I immediately went back to my other major project, the collaboration with Ludovic and Alessio. Here again, just when the proof seemed to be in jeopardy, we were able to overcome all the obstacles facing us and everything began to click, as if by magic. Our good fortune included one true miracle, by the way, an enormous calculation in which fifteen terms recombined to constitute a perfect square—an outcome as unhoped for as it was unexpected, since ultimately what we succeeded in demonstrating was the opposite of what we had set out to demonstrate!

With regard to Landau damping, it must be admitted that Clément and I didn't manage to solve quite everything. For electrostatic and gravitational interactions, the most interesting cases, we were able to show that damping occurs on a gigantic time scale, but not an infinitely long one. And since we were stymied on this point, we were also stymied on regularity—we couldn't find a way to get out from the analytic framework. Very often at the end of my talks somebody would ask one of these two questions: *In the case of Coulomb or Newton interaction, does one also have damping in infinite time? Can one do without the analyticity assumption?* In either case my response was the same, that I couldn't say anything without consulting my lawyer first. Honestly, I don't know whether there's a profound mystery here, or whether we simply weren't clever enough to work out the answer.

Oh, here's a young woman walking alone, like me, one of the participants in the mathematics program. It turns out she heard my talk on optimal transport and found it a good introduction to the subject. We walk together the rest of the way back to Simonyi Hall, talking of mathematics in the fading spring light of a Princeton evening.

Need to go up to my office one last time. It's empty—except for an enormous stack of handwritten notes and rough drafts containing all my attempts to deal with one problem or another, successful and aborted ones alike, together with all the intermediate versions I had carefully composed, then compulsively printed out and furiously corrected.

I would love to take them back to France with me, but it's just too much to deal with at this point. We've got more than enough luggage to check at the airport as it is! There's no choice but to throw it all out. . . .

Seeing me standing there, staring at this huge pile of paper, my young companion immediately comprehends the minor tragedy of having to do away with the work of so many months, all of it fraught with so much emotion. She kindly offers to help me stuff it into the wastepaper basket.

We stuff it into the basket—and pile the rest all around on the floor. There was enough to fill at least four baskets!

My stay in Princeton is now well and truly at an end.

•

WELCOMING REMARKS BY THE DIRECTOR OF THE INSTITUT HENRI POINCARÉ, NOVEMBER 6, 2009

For a long time I found the idea of bringing together young female mathematicians puzzling—until earlier this year when I participated myself, as a speaker, in the annual Program for Women and Mathematics at the Institute for Advanced Study in Princeton. The liveliness and enthusiasm of everyone who took part made an unforgettable impression on me. It is my sincere hope that the atmosphere of the lectures and discussions awaiting you in this ninth "Forum des Jeunes Mathématiciennes" at the Institut Henri Poincaré will be no less relaxed, and no less intense, than the one it was my good fortune to discover for the first time a few months ago. Welcome to the Maison des Mathématiciennes!

THIRTY-THREE

How strange to be back home again, after being away so long!

You haven't really settled in after an extended stay abroad until you've gone out food shopping. You're reunited at long last with your favorite grocers, you find the breads and cheeses you've been missing, you're astonished to hear everyone speaking French. A glass of raw milk, my first in six months, brought tears to my eyes. The ciabatta's soft crumb and the baguette's crispy crust require no comment.

But now that I'm back in my element, nothing is quite the same anymore. Remodelers were hard at work while we were gone, you can scarcely recognize the apartment. . . . But that's nothing compared to the inner transformation I've experienced as a result of the work I did in Princeton. I feel like a mountain climber who has come back down to earth, his head filled with vivid images of the heights he has scaled. Chance redirected the course of my research to a degree I wouldn't have thought possible six months ago.

In the 1950s, the Metropolis–Hastings algorithm brought about a scientific revolution by showing that, if a space is too rich in possibilities, you are often better off moving around at random rather than exploring it in a systematic way or selecting probability distribution samples in a perfectly haphazard manner. Since then the algorithm has spawned an entire field of research, so-called Markov chain

Monte Carlo (MCMC) methods, whose unreasonable effectiveness in physics, chemistry, and biology has yet to be satisfactorily explained. Exploration in the MCMC sense is neither wholly deterministic nor completely aleatory; it is something in between, a compromise between choosing the best possible path at every step and choosing each successive step at random. That way you are free from time to time to choose a path that, although it may not seem the best one at first, promises to lead to increasingly better options down the road.

But there's really nothing new about this. We do the same thing in everyday life: passing more or less at random from one situation to another enables us to explore many more possibilities than we could otherwise. The world of mathematics is no different: unpredictable encounters lead us from one problem to another, sometimes even from one scientific continent to another.

And now, just when everything has finally been put back in its place, a new home is waiting to be furnished. My personal belongings are already packed in boxes; soon the movers will come and take away all the things I've lived with for so many years. The futon that my mother was amazed anyone would want to sleep on (like reinforced concrete, she said). The music system that after fifteen years of intensive use has more than made good on its promise of high fidelity. The hundreds of CDs that nearly ate up my salary as an undergraduate research assistant, the scavenged cassettes and the second-hand vinyl records. The massive wooden partners desk, the great colonial bookcases that hold vast numbers of books, the heavy armchair carved from a single piece of solid wood that I found in London, sculptures bought in the Drôme, my grandfather's paintings—all these things are going to accompany me on my new adventure. Three days from now I take up my duties as director of the Institut Henri Poincaré in Paris: my predecessor vacates his office on June 30, I move in July 1. No time to prepare; I'll just have to learn the job as I go along. The beginning of a new period in my life.

A new stage in my personal MCMC.

•

Having survived a difficult period in the 1970s and 1980s, the "Home of Mathematics" embarked on a new chapter of its history in 1990. The substantial costs of the Institute's physical renovation were mainly assumed by the state, under the terms of a four-year agreement with the Université Pierre et Marie Curie, the governing body of the new IHP, with additional funding from the Centre National de Recherche Scientifique.

The new administrative structure was put in place under the guidance of the mathematician Pierre Grisvard, who died prematurely in 1994, a few months before the Institute's official reinauguration by the minister for higher education and research. Grisvard was succeeded as director by Joseph Oesterlé (Université Pierre et Marie Curie), who was followed by Michel Broué (Université Denis Diderot) in 1999, and then by Cédric Villani (École Normale Supérieure de Lyon) in 2009.

[From a brief history of the Institut Henri Poincaré]

THIRTY-FOUR

Prague! If ever there was a mythic city in Europe, this is it. The legend of the Golem, Messia's song, Mairowitz's biography of Kafka—thoughts of these and many more things filled my head as I strolled through streets with thousand-year-old clocks and strip clubs cheek by jowl, and dance clubs with lines of students wearing devil horns and superhero capes waiting to get in.

A few weeks ago, in the little town of Oberwolfach, passersby stared wide-eyed at my costume, but in Prague I could almost pass for an accountant.

Yesterday was the opening of the International Congress of Mathematical Physics, held every three years under the auspices of the international association of the same name, the IAMP. I was one of four who were honored with great pomp and ceremony as winners of the Henri Poincaré Prize. In addition to the Austrian Robert Seiringer (recognized, as I was, in the junior category), the laureates included the Swiss Jürg Frölich and the Russian Yasha Sinai. Specialists in classical and quantum mechanics, statistical physics, and dynamical systems—all of them friends of the farsighted Joel Lebowitz, who long ago invited them to serve on the editorial board of his *Journal of Statistical Physics*. I was both pleased and flattered to find myself in such good company.

Winning the Poincaré Prize meant I was entitled to give a plenary address to the Congress even though I hadn't been selected as a speaker beforehand. Although I had received the prize for my work on Boltzmann, I chose instead to talk about Landau damping: an unexpected opportunity to communicate my results to the most distinguished audience of mathematical physicists imaginable.

Three minutes before I was scheduled to begin, my heart was beating wildly, adrenaline coursing through my veins. But once I began to speak, I felt calm and sure of myself.

"It so happens that I was recently appointed director of the Institut Henri Poincaré, and now I have been awarded the Henri Poincaré Prize. Just a coincidence, of course—but I like it. . . ."

I had meticulously prepared my speech. Everything went as planned, I finished right on time.

"To conclude, let me note a nice coincidence. In order to treat the singularity of the Newton interaction, you use the full power of the Newton scheme. Newton would be delighted! Again, this is just a coincidence—but I like it. . . ."

Thunderous applause. On certain faces in the audience I thought I detected looks not only of astonishment and admiration, but also, to some extent, of fear. To be honest, even I find the proof intimidating!

And then there were the Czech girls. Before my turn came to speak they didn't pay much attention to me, but afterward it was an entirely different story: they crowded around, congratulated me on the clarity of my exposition; one young woman rather emotionally recited a little speech in shaky French.

Unsurprisingly, my results once more prompted the same two questions. Can the analytic regularity assumption be relaxed? For a Newton interaction, isn't it possible for the damping to go on and on, in infinite time? None of this mattered in the least to my Portuguese friend Jean-Claude Zambrini. Amid the clamor outside the lecture hall he exclaimed, "Since you attract coincidences, Cédric, let's hope that you'll receive an invitation from the Fields Institute next!"

The Fields Institute for Research in Mathematical Sciences— which plays no role in the awarding of the Fields Medal—is located in Toronto, and regularly sponsors colloquia on all sorts of topics.

Jean-Claude and I had a good laugh. But only a month and a half later, by pure coincidence, the invitation arrived.

•

Date: Tue, 22 Sep 2009 16:10:51 -0400 (EDT)
From: Robert McCann <mccann@math.toronto.edu>
To: Cedric Villani <Cedric.VILLANI@umpa.ens-lyon.fr>
Subject: Fields 2010

Dear Cedric,

Next fall I am involved in organizing a workshop on "Geometric Probability and Optimal Transportation" Nov 1-5 as part of the Fields Theme Semester on "Asymptotic Geometric Analysis".
You will certainly be invited to the workshop, with all expenses covered, and I hope you will be able to come. However, I also wanted to check whether there is a possibility you might be interested in visiting Toronto and the Fields Institute for a longer period, in which case we would try to make the opportunity attractive.
Please let me know,

Robert

THIRTY-FIVE

Back home, about a week ago, the children made the acquaintance of a baby wild boar their uncle had captured with his bare hands. How I wish I could have been there!

But I'd convinced myself that during their fall break from school my time would be better spent lecturing in the United States—a grueling tour that will take me through half the country in just ten days. Already I've been to Cambridge, paying visits to MIT (in the footsteps of Wiener and Nash!) and Harvard. Now I'm in New York. I console myself with the thought that once I return to France I'll be able to go see the little piglet and take him for a walk in the woods.

On opening my email this evening, my heart skipped a beat: a message from *Acta Mathematica*, considered by many to be the most prestigious of all mathematical research journals. This is where Clément and I submitted our one-hundred-eighty-page monster for publication. No doubt that's what they're writing me about.

But . . . it's not even four months since we submitted it! Considering the size of the manuscript, it's much too soon for the editorial board to have decided in our favor. Only one possible explanation: they're writing to say that the article has been rejected.

I open the message, skim the cover letter, feverishly scroll through the attached referees' reports. Lips pursed, I read through them again, this time more carefully. Six reports, very positive on the whole. Not a word to be said against us, except . . . sure enough, always the same misgiving: it's the analyticity assumption that bothers them—that and the limit case in large time. Always the same two questions. I've already had to respond to them dozens of times over the past few months—and now it looks like they've sunk our hopes of publication! The editor who signed the letter isn't persuaded that our results are definitive, and because the manuscript is so long he feels obliged to hold it to a stricter standard of review than is customary.

But this simply isn't right!! After all the ingenuity we've displayed, all the new ground we've broken, all the technical obstacles we've managed to overcome, all the days of hard work that stretched on late into the night, night after night—and *still* it's not good enough for them?? Give me a *break*, for Chrissake!

But then . . . I noticed another message, informing me that I'd just won the Fermat Prize. Named after the great Pierre de Fermat, whose mathematical enigmas infuriated the whole of seventeenth-century Europe; the prince of amateurs, as he was known, who revolutionized number theory and the calculus of variations, and laid the foundations of modern probability theory. Since 1989 the prize established in his name has been awarded every two years to one or two mathematicians under the age of forty-five who have made major contributions in one of these domains.

The news comes as a great consolation. Even so, it isn't enough for me to get over the frustration of seeing our article turned down. To get over that, I'd need a big long hug. At a minimum.

•

In 1882, Gösta Mittag-Leffler persuaded a group of fellow mathematicians in his native Sweden and the other Scandinavian countries to create a review devoted to publishing research of the highest quality. It was named Acta Mathematica, *with Mittag-Leffler as its first editor-in-chief.*

Mittag-Leffler was in touch with the top mathematicians in the world. Blessed with unfailingly good taste, and an unusually large appetite for risk, he rapidly succeeded in attracting the most exciting new work of the day. His favorite author was the brilliant and unpredictable Henri Poincaré, whose long, trailblazing manuscripts Mittag-Leffler published without the least hesitation.

The most famous episode in the history of Acta Mathematica *is also one of the most famous episodes in Poincaré's career. In 1887, on Mittag-Leffler's advice, King Oscar II of Sweden announced an international mathematics competition. Poincaré accepted the challenge, and from a short list of topics selected the stability of the solar system, an open problem since Isaac Newton first posed it two hundred years earlier! Newton had written down the basic equations for planetary movement in the solar system (the planets are attracted by the sun and attract one another), but he was unable to show that these equations entail the stability of the solar system itself, or to determine whether, on the contrary, they contain the seeds of a preordained catastrophe—the collision of two planets, for example. Every mathematical physicist of Poincaré's time was familiar with the problem.*

Newton thought that the solar system was inherently unstable, and that the stability we perceive is due to a divine helping hand. Later, however, the results obtained first by Laplace and Lagrange, and then by Gauss, made it clear that Newton's system is stable over a huge period of time, perhaps as long as a million years—much longer than Newton himself believed. This meant that the behavior of the sun and the planets had been qualitatively predicted on a time scale far greater than the whole of recorded human history!

One question nonetheless remained unanswered: Once such a huge period of time has passed, is catastrophe likely—even inevitable? If we wait not one million but one hundred million years, are Mars and Earth in real danger of colliding? This turned out to be no ordinary problem, for it concealed fundamental questions about physics itself.

Poincaré didn't propose to treat the motions of the entire solar system. Way too complicated! Instead he considered an idealized model of the system on a reduced scale, taking into account only two bodies turning around the sun, one of which he assumed to be minuscule by comparison with the other—rather as though one were to neglect all the planets except Jupiter and Earth. Poincaré studied this simplified problem, and then simplified it still further, until he had finally gotten to the very heart of the matter. Devising novel methods to suit his purpose, he proved the eternal stability of this scaled-down system!

For this feat he earned the glory of King Oscar's prize and the money that came with it.

It was understood that the winning paper would appear in Acta Mathematica. *The assistant editor responsible for preparing the text for the printer noticed a few rather doubtful things about Poincaré's solution, but this hardly came as a surprise: everyone knew that Poincaré wasn't a model of clarity. The editor duly transmitted his queries to the author, with all the deference due to one of the towering figures of the age.*

By the time Poincaré realized that a serious error had crept into the proof, the issue containing his article had already been published! An erratum *would not suffice, because the results themselves were fatally contaminated.*

Mittag-Leffler was unflustered. He managed on various pretexts to get all the copies back, one after another, before anyone spotted the error. Poincaré bore the full cost of pulping the original print run—a sum that exceeded the value of King Oscar's award!

At this point the story takes an extraordinary turn. Poincaré redeemed his mistake by seizing the opportunity it presented to create a whole new field of scientific investigation. Having remedied the defects of his argument and revised its conclusions accordingly, he found that he had in fact shown the opposite of what he felt sure he had demonstrated: instability was possible after all!

In the corrected and expanded form in which it was republished the following year, in 1890, the article gave rise to the study of dynamical systems, a topic that more than a century later keeps thousands of scientists busy in fields ranging from physics to economics. Quasi-periodic orbits, chaos theory, sensitive dependence on initial conditions, fractals—all these things are found in embryonic form in Poincaré's revised paper. What could have been a disaster for Acta Mathematica *turned out to be a triumph.*

The journal's renown continued to grow, and today it is one of the most prestigious scientific journals in the world, perhaps even the most prestigious of all. To see an article you've written take up some of the six hundred pages Acta Mathematica *publishes each year is almost enough to assure your future as a professional mathematician.*

When Poincaré died, in 1912, he was eulogized in France as a national hero. Four years later, in 1916, Mittag-Leffler's house outside Stockholm was converted into an international research center where mathematicians from all over the world could gather to ponder old problems and work together on new ones. The Mittag-Leffler Institute, the very first of its kind, is still going strong today. In 1928, the year after Gösta Mittag-Leffler himself passed away, a second such center was founded in Paris, based on the same principles of international fellowship, with special emphasis on postgraduate training. Appropriately, it was named after Henri Poincaré.

Henri Poincaré

Gösta Mittag-Leffler

THIRTY-SIX

In my hotel room in Ann Arbor. I'm here for a few days at the University of Michigan—a great university with some mathematicians of the first rank.

Clément was really demoralized by *Acta Mathematica*'s rejection. He wanted to try to convince them to reconsider their decision, to make them understand why our result is so innovative and so important, even if there's still a little gray area. . . .

But I've had more experience dealing with these people than he has. And as an editor of a rival journal, *Inventiones Mathematicae*, I know firsthand how tough one must be in judging manuscripts submitted for review. The *Acta* editors are even thicker-skinned than I am. Nothing will make them change their minds unless a referee can be shown to have acted in bad faith (no hint of this in our case) or unless the proof can be strengthened.

One possibility would be to cut the beast in two in the hope of improving our chances of publishing elsewhere, but I cannot bring myself to do such a thing. . . . So we're going to sleep on it.

My talks here have gone well so far, but the same questions keep coming up over and over again. I've discussed the matter with Jeff Rauch, a leading expert on partial differential equations who has spent a lot of time working in France. Jeff wasn't shocked by the fact that the result doesn't hold for infinite time, but he didn't like the analyticity assumption. There are others, of course, who would like to see it work in infinite time and who don't mind the assumption. I could tell myself that it's no big deal; but since I trust Jeff's judgment, I find his reservations troubling. So this evening I decided to sit down and write out an argument that would convince him that our proof is as fine a thing as human minds can make it, and that very few, if any, improvements are possible. A good bit of work, but it's as much for my benefit as for his.

Jeff Rauch

The time went by. Sitting on the bed, scribbling away. I couldn't find a way to convince myself. . . . And if I can't find a way to convince myself, there's very little chance that I'll have any luck convincing Jeff!!

What if I went astray somehow, if my estimates weren't precise enough? Here, though, I haven't lost anything . . . there, it would really *be surprising if I got it wrong . . . here, it's optimal . . . and there, well, simplifying can only help—unless I've been utterly bewitched. . . .*

Like a cyclist inspecting his chain for a weak link, I carefully went through the proof, checking the soundness of the argument at each step.

And there . . .

There! *That's where I may have been too careless! Jesus, how could I have not seen that the modes are diverging—and that my comparison by summation was too rough!?!? If it's the sup in relation to the sum, obviously I'm going to lose something!! Well, okay, it got swamped by all the technical detail. . . .*

No time for grumbling, I had to work out the implications.

But of course . . . the modes diverge, the weight shifts, if you put them all together the loss is huge!! That means they've got to be kept separate!!!

This was my moment of illumination, the moment when the light bulb went on in my head. I jumped up from the bed and feverishly paced up and down, pencil in one hand, a sheet of paper covered with cabalistic formulas in the other. I kept on staring at them. This time it wasn't a question of fixing an error; it was a question of improving our results.

How are we going to get from here to there?

I don't know. But now at last we're on our way. We're going to take the whole thing apart and put it back together. We've finally figured out how to lay these two infernal objections to rest, once and for all.

·

Since $\gamma = 1$ is the most interesting case, it is tempting to believe that we stumbled on some deep difficulty. But this is a trap: a much more precise estimate can be obtained by *separating modes* and estimating them one by one, rather than seeking an estimate on the whole norm. Namely, if we set

$$\varphi_k(t) = e^{2\pi(\lambda t + \mu)|k|} |\hat{\rho}(t, k)|,$$

then we have a system of the form

$$\varphi_k(t) \le a_k(t) + \frac{ct}{(k+1)^{\gamma+1}} \varphi_{k+1}\left(\frac{kt}{k+1}\right). \tag{7.15}$$

Let us assume that $a_k(t) = O(e^{-ak} e^{-2\pi\lambda|k|t})$. First we simplify the time dependence by letting

$$A_k(t) = a_k(t)e^{2\pi\lambda|k|t}, \quad \Phi_k(t) = \varphi_k(t)e^{2\pi\lambda|k|t}.$$

Then (7.15) becomes

$$\Phi_k(t) \leq A_k(t) + \frac{ct}{(k+1)^{\gamma+1}}\Phi_{k+1}\left(\frac{kt}{k+1}\right). \qquad (7.16)$$

(The exponential for the last term is right because $(k+1)$ $(kt/(k+1)) = kt$.) Now if we get a subexponential estimate on $\Phi_k(t)$, this will imply an exponential decay for $\varphi_k(t)$.

Once again we look for a power series, assuming that A_k is constant in time, decaying like e^{-ak} as $k \to \infty$; so we make the ansatz $\Phi_k(t) = \sum_m a_{k,m} t^m$ with $a_{k,0} = e^{-ak}$. As an exercise, the reader can work out the doubly recurrent estimate on the coefficients $a_{k,m}$ and deduce

$$a_{k,m} \leq \text{const.}\ A(ke^{-ak})\, k^m\, c^m\, \frac{e^{-am}}{(m!)^{\gamma+2}},$$

whence

$$\Phi_k(t) \leq \text{const.}\ A e^{(1-\alpha)(ckt)^\alpha}, \quad \forall\, \alpha < \frac{1}{\gamma+2}. \qquad (7.17)$$

This is subexponential even for $\gamma = 1$: in fact, we have taken advantage of the fact that *echoes at different values of* k *are asymptotically rather well separated in time.*

Therefore, as an effect of the singularity of the interaction, *we expect to lose a fractional exponential on the convergence rate*: if the mode k of the source decays like $e^{-2\pi\lambda|k|t}$, then φ_k, the mode k of the solution, should decay like

$e^{-2\pi\lambda|k|t}\,e^{(c|k|t)^{\alpha}}$. More generally, if the mode k decays like $A(kt)$, one expects that $\phi_k(t)$ decays like $A(kt)\,e^{(c|k|t)^{\alpha}}$. Then we conclude as before by absorbing the fractional exponential in a very slow exponential, at the price of a *very large* constant: say,

$$e^{t^{\alpha}} \leq \exp\left(c\varepsilon^{-\frac{\alpha}{1-\alpha}}\right)e^{\varepsilon t}.$$

[From my 2010 Luminy Summer School lecture
notes on Landau damping]

THIRTY-SEVEN

Charlotte
November 1, 2009

In transit between Palm Beach and Providence. An impersonal airport in North Carolina; it could be anywhere, really. Removing all metal items when you pass through the security checkpoint is a bit of a chore for most people, but when you also wear cuff links and a fob watch, and on top of that carry one or two USB flash drives and a half dozen pens in your pockets . . .

In Florida, at the international workshop on geometric inequalities that Emanuel Milman helped organize in Boca Raton, the living was easy! From downtown it was only a few steps to the beach. And the ocean was like a warm bath. Even at night the temperature was ideal, no one around, no need for a bathing suit . . . really, it was like swimming in a bathtub—a bathtub with tides and soft sand! And all this in November!

But now the fun's over, back to the cold. It's going to happen so fast, thanks to the unnatural swiftness of modern air travel!

In Boca Raton I was able to forget Landau damping for a day or two, but now once again it occupies my every waking moment. I'm beginning to see what needs to be done to make the proof stronger overall, how to exploit the flash of inspiration that struck me in Ann Arbor. Still, it's going to be a huge amount of work! Waiting for the connecting flight to Providence, I wasn't sure if I really had enough confidence to talk about the new plan of attack at Brown without having yet worked out all the details. It's a very important talk: Yan Guo, the one by whom the Problem came, is going to be in the audience.

I took out a blank piece of paper and began to sketch the new plan, redoing calculations etc., when all of a sudden it leapt to my eyes—there's something wrong, a contradiction.

I can't possibly prove an estimate that sharp. . . .

After a few minutes I'd convinced myself that there must be a mistake somewhere in one of the especially involved parts of the proof. Or could it be that—it's *all* wrong? The airport began to pitch and roll around me. . . .

I pulled myself together.

Cédric, the error can't be very serious. Everything else holds together too well. The error has to be local, it's got to be right here, in this passage somewhere. The reason for it has to be that the calculation is obscured by these two miserable little shifts—the double time-shift you introduced after coming back from the museum!

But Clément definitely showed that we can do without them!! So be it, we'll have to get rid of them after all, they're too dangerous. In a proof this complex, the least source of obscurity must be ruthlessly eliminated.

Even so, if I hadn't come up with this double shift we might have been stymied for good. The double shift was the thing that gave us hope, that allowed us to move forward again. Later on, we saw that it wasn't necessary. And so what if in the end it turns out to be wrong!? Fine. We'll rewrite everything, without even mentioning it.

For the moment I need to decide what to say tomorrow at Brown. I'll have to say that I've found a way to improve the proof, and then explain why it's important, because it will answer the two criticisms of our result that have been made over and over again. But I mustn't cheat—no bluffing this time!

Palm Beach to Providence, more turbulence than expected . . .

•

SUMMARY OF YOUR WEST PALM
BEACH–PROVIDENCE TRIP

Flight details: Sunday 1 November 2009
    Travel time: 6 hrs 39 mins
Depart: 03:00 PM, West Palm Beach, FL (PBI)
Arrive: 04:53 PM, Charlotte, NC (CLT)
    US Airways 1476 Boeing 737-400 Economy Class

Depart: 07:49 PM, Charlotte, NC (CLT)
Arrive: 09:39 PM, Providence, RI (PVD)
    US Airways 828 Airbus A319 Economy Class

•

Coulomb/Newton (most interesting case)
Coulomb/Newton interaction and analytic regularity are *both* critical in the proof of nonlinear Landau damping; but the proof still works for *exponentially large times* "because"

- the expected linear decay is exponential
- the expected nonlinear growth is exponential
- the Newton scheme converges bi-exponentially

Still, it seems possible to go further by exploiting the fact that *echoes at different spatial frequencies are asymptotically rather well separated.*

        [From notes for my talk at Brown University,
                       November 2, 2009]

THIRTY-EIGHT

Saint-Rémy-lès-Cheuvreuse
November 29, 2009

Sunday morning, scribbling away in bed. One of the special moments in the life of a mathematician.

I'm rereading the final version of our article, crossing out, correcting. More relaxed than I've been in months! We've rewritten the whole thing. Completely eliminated the treacherous double shift. Succeeded in exploiting the asymptotic time separation of the echoes, recast the main part of the proof, substituted a mode-by-mode study for the original aggregate-level treatment, relaxed the analyticity condition, and, last but not least, included the Coulomb case in infinite time, the thing that everyone had been complaining about for so long. . . . Everything revised, everything simplified, everything checked once more, everything improved, everything gone over one last time.

All of this could easily have taken three months, but in our present state of feverish excitement three weeks turned out to be enough.

More than once, going through the argument with a fine-toothed comb, we've wondered how in the world we could have come up with this little trick or that little piece of cleverness.

The result is now much stronger. In the process we also managed to solve a problem that has long intrigued specialists such as Guo, technically known as the orbital stability of homogeneous nonmonotonic linearly stable equilibria.

We added a few passages, but simplifying has reduced the

length elsewhere, so that now the manuscript is scarcely longer than the one we submitted earlier.

New computer simulations came in as well. When I saw the first batch of results last week, I was staggered: the numerical calculations that Francis had performed using an extremely precise formula seemed to completely contradict our theoretical results! But I didn't buckle. I told Francis I was skeptical, and he reran everything using another method that is considered to be even more accurate. When the second batch came in, the results were consistent with the theoretical prediction. Phew! Just goes to show that computer simulations are no substitute for qualitative insight.

Tomorrow we'll be ready to make the new version available via the Internet. And at the end of the week we'll be able to resubmit the paper to *Acta Mathematica*, with a much greater chance of success this time around.

In my idle moments I can't help but think of Poincaré himself. One of his most famous articles was, well, not rejected by *Acta*, but nevertheless withdrawn, and then corrected and finally republished. Perhaps the same thing is going to happen to me? It's already been my Poincaré year: I won the Henri Poincaré Prize, and I'm head of the Institut Henri Poincaré.

Poincaré . . . careful, Cédric, beware delusions of grandeur!

•

Paris, 6 December 2009

From: Cédric Villani
École Normale Supérieure de Lyon
    & Institut Henri Poincaré
11, rue Pierre & Marie Curie
F-75005 Paris
FRANCE
cvillani@umpa.ens-lyon.fr

To: Johannes Sjöstrand
Editor, *Acta Mathematica*
IMB, Université de Bourgogne
9, avenue A. Savarey, BP 47870
F-21078 Dijon
FRANCE
johannes.sjostrand@u-bourgogne.fr

Re: Resubmission to *Acta Mathematica*

Dear Professor Sjöstrand:

Following your letter of October 23, we are glad to submit a new version of our paper, "On Landau damping," for possible publication in *Acta Mathematica*.

We have taken good note of the concerns expressed by some of the experts in the screening reports on our first submission. We believe that these concerns are fully addressed by the present, notably improved, version.

First, and maybe most importantly, the main result now covers Coulomb and Newton potentials; in an analytic setting this was the only remaining gap in our analysis.

Analyticity is a classical assumption in the study of Landau damping, both in physics and mathematics; it is mandatory for exponential convergence. On the other hand, it is very rigid, and one of the referees complained that our results were tied to analyticity. With this new version this is not so, since we are now able to cover some classes of Gevrey data.

In the first version, we wrote: "[W]e claim that unless some new stability effect is identified, there is no reason to believe in nonlinear Landau damping for, say, gravitational interaction, in any regularity class lower than analytic." Since then we have identified precisely such an effect (echoes occurring at different frequencies are asymptotically well separated). Exploiting it led to the above-mentioned improvements.

As a corollary, our work now includes new results of stability for homogeneous equilibria of the Vlasov–Poisson equation, such as the stability of certain nonmonotonic distributions in the repulsive case (a longstanding open problem), and stability below the Jeans length in the attractive case.

Another referee expressed a reservation about our use of nonconventional functional spaces. While this may be the case for our "working norm," it is not so for the naïve norm appearing in our assumptions and conclusions, already used by others. Passing from one norm to another is done by means of Theorem 4.20.

The paper was entirely rewritten to incorporate these improvements, and carefully proofread. To prevent any further increase in the paper's length, we have cut all expository passages and comments which were not strictly related to our main result; most of the remaining remarks are intended simply to explain the results and methods.

A final comment about length: we are open to discussion regarding adjustments to the organization of the paper, and we note that the modular presentation of the tools used in our work probably makes it possible for some referees to work as a team, thereby hopefully alleviating their task.

We very much hope that this paper will satisfy the experts and remain

Yours truly,

Clément Mouhot & Cédric Villani

•

ON LANDAU DAMPING

C. MOUHOT AND C. VILLANI

ABSTRACT. Going beyond the linearized study has been a longstanding problem in the theory of Landau damping. In this paper we establish exponential Landau damping in analytic regularity. The damping phenomenon is reinterpreted in terms of transfer of regularity between kinetic and spatial variables, rather than exchanges of energy; phase mixing is the driving mechanism. The analysis involves new families of analytic norms, measuring regularity by comparison with solutions of the free transport equation; new functional inequalities; a control of nonlinear echoes; sharp scattering estimates; and a Newton approximation scheme. Our results hold for any potential no more singular than Coulomb or Newton interaction; the limit cases are included with specific technical effort. As a side result, the stability of homogeneous equilibria of the nonlinear Vlasov equation is established under sharp assumptions. We point out the strong analogy with the KAM theory, and discuss physical implications.

CONTENTS

Keywords. Landau damping; plasma physics; galactic dynamics; Vlasov-Poisson equation.

AMS Subject Classification. 82C99 (85A05, 82D10)

THIRTY-NINE

Reading email first thing in the morning, as soon as you've gotten out of bed, is a sort of intellectual drug injection. Not too much of a jolt, just enough to get you started

Among the new messages today, Laurent Desvillettes sends unhappy news: our mutual friend Carlo Cercignani has died.

Carlo's name is inseparable from that of Ludwig Boltzmann. Carlo devoted his professional life to Boltzmann, to Boltzmann's theories, to his celebrated equation and its many applications. He wrote three of the standard works on Boltzmann, one of them the first research monograph I ever read.

Carlo's mathematical interests were nevertheless extraordinarily diverse. Boltzmann allowed him to explore a great many topics, some of them having no obvious connection to the equation Carlo so cherished.

And that wasn't the half of this universal man, polyglot and polymath, highly cultured, who refused to limit himself to the sciences: his works include a play for the theater, a collection of poems, and translations from Homer.

My first important result, or at least the first one I was really proud of, concerned what is known as Cercignani's conjecture. I was twenty-four years old and raring to go when Giuseppe Toscani invited me to spend two weeks at the University of Pavia in October 1997. An idea had just occurred to Giuseppe about how to tackle this famous conjecture, and he suggested that I give it a go during my brief stay. After a few hours I could see that his crude frontal assault had no chance of success . . . but I did notice an interesting calculation, something that had the ring of truth about it, the hint of a remarkable new identity. It was all I needed: the mathematical rocket was ready to be launched.

I began by showing that Cercignani's conjecture regarding entropy production in Boltzmann's equation could be reduced to an estimate of entropy production in a problem in plasma physics that, as it happened, I had already studied with Laurent. And then I mixed in a bit of information theory, something that has always fascinated me. An incredible combination of circumstances that never would have come about had Giuseppe's misguided intuition not struck him at precisely the moment I turned up!

Working together, we had *almost* cracked the case by the time I had to leave. Later the same month the opportunity presented itself to announce our results to an audience of the leading experts on the Boltzmann equation at a conference in Toulouse. It was here that Carlo, like many others, first came to know about me. I can still hear the excitement in his voice as he urged me on: "Cédric, prove my conjecture!"

At twenty-four, it was one of my first published articles. Five years later, in my twenty-fourth article, I returned to the problem, only now with more experience and more technique, and succeeded finally in proving Carlo's brilliant conjecture. He was so proud of me.

The Boltzmann equation still has quite a few loose ends, however, and Carlo was counting on me to tie up some of the most important—and the most maddening—ones. That was one of my ambitions as well. But then, without warning, I wandered off, first in the direction of optimal transport and geometry, then of the Vlasov equation and Landau damping.

I still have it in mind to come back to Boltzmann, just not now. But even if one day my dream comes true, I shall never know the joy of telling Carlo that I've tamed his favorite monster, the one he loved before all others.

•

Cercignani's conjecture concerns the relationship between entropy and entropy production in a gas. Let's simplify, and leave to one side the spatially inhomogeneous character of the gas, so that the only thing that matters is the velocity distribution. Therefore, let f(v) be a velocity distribution in a gas away from equilibrium: since this distribution is not equal to the Gaussian γ(v), *the entropy is not as high as it might be. Boltzmann's equation predicts that the entropy will increase. Will it increase by a lot or only by very little?*

Cercignani's conjecture suggests that the instantaneous increase in entropy is at least proportional to the difference between the entropy of the Gaussian and the entropy of the distribution that we're interested in:

$$\dot{S} \geq K\big[S(\gamma) - S(f)\big].$$

The conjecture has implications for figuring out how fast the distribution converges to equilibrium—a fundamental question connected with Boltzmann's fascinating discovery of irreversibility.

Laurent Desvillettes worked on Cercignani's conjecture in the early 1990s, and after him Eric Carlen and Maria Carvalho. They obtained partial results that opened up completely new perspectives, but they were still a long way from proving it. Cercignani himself, with the help of the Russian mathematician Sasha Bobylev, had shown that the conjecture was overly optimistic, that it couldn't be true . . . unless perhaps if one were to assume extremely strong collisions, interactions harder than those of so-called hard spheres, with the cross-section increasing at least proportionally to the relative velocity—"very hard spheres," in the jargon of the kinetic theory of gases.

But in 1997 Giuseppe Toscani and I demonstrated the existence of a bound that is "almost" as good:

$$\dot{S} \geq K_{\varepsilon}\big[S(\gamma) - S(f)\big]^{1+\varepsilon},$$

where ε is as small as one likes, under certain quite restrictive conditions governing collisions.

In 2003, I showed that this result holds true for all interactions satisfying certain reasonable assumptions; and, more strikingly still, I managed to show that the conjecture is true if the high-velocity collisions are of the very hard sphere type. The key identity, discovered with Toscani six years earlier, was the following:

If $(S_t)_{t \geq 0}$ is the semigroup associated with the Fokker–Planck equation, $\partial_t f = \nabla_v \cdot (\nabla_v f + fv)$, and $\varepsilon(F, G) := (F - G) \log(F/G)$, then

$$\frac{d}{dt}\bigg|_{t=0} [S_t, \varepsilon] = -\mathcal{J},$$

$$where \quad \mathcal{J}(F, G) = \big|\nabla \log F - \nabla \log G\big|^2 (F + G).$$

This identity plays a key role in the representation formula

$$\dot{S}(f) \geq K \int_0^{+\infty} e^{-4Nt} \int_{\mathbb{R}^{2N}} (1 + |v - v_*|^2)$$

$$\times \mathcal{J}(S_t F, S_t G) \, dv \, dv_* \, dt,$$

Carlo Cercignani

where $F(v, v_*) = f(v)f(v_*)$ and $G(v, v_*)$ is the average of all the products $f(v')$ $f(v'_*)$ when (v', v'_*) describes all pairs of postcollision velocities compatible with the precollision velocities (v, v_*). This formula lies at the heart of the solution of Cercignani's conjecture.

Theorem *(Villani, 2003). Let $S(f) = -\int f \log f$ denote the Boltzmann entropy associated with a velocity distribution $f = f(v)$. Let B be a Boltzmann collision kernel satisfying $B(v - v_*, \sigma) \geq K_B(1 + |v - v_*|^2)$ for some constant $K_B > 0$, and denote by $\dot{S}$ the associated entropy production functional,*

$$\dot{S} = \frac{1}{4} \iiint \left(f(v')f(v'_*) - f(v)f(v_*) \right)$$

$$\times \log \frac{f'(v)f'(v_*)}{f(v)f(v_*)} \, B \, dv \, dv_* \, d\sigma.$$

Let $f = f(v)$ be a probability distribution on $\mathbb{R}^N$ with zero mean and unit temperature. Then

$$\dot{S}(f) \geq \left(\frac{K_B |S^{N-1}|}{4(2N+1)} \right)(N - T^*(f))[S(\gamma) - S(f)],$$

where

$$T^*(f) = \max_{e \in S^{N-1}} \int_{\mathbb{R}^N} f(v)(v \cdot e)^2 \, dv.$$

FORTY

Late afternoon in my spacious office at the Institut Henri Poincaré. I had the handsome blackboard enlarged and got rid of a few pieces of furniture to make more room. I've given a lot of thought to how I want to redecorate.

First, the bulky air conditioning unit has got to go—it's normal to be hot in the summer!

Against the wall, a large display cabinet will hold some personal items and a few of the jewels from the Institute's collection of geometric models.

To the left of my desk I plan to install the bust of a rather stern-looking Henri Poincaré that his grandson, François Poincaré, generously donated to the IHP.

And behind me a large space is reserved for a portrait of Catherine Ribeiro! I've already selected the image, found it on the Internet: Catherine with her arms spread wide in a vast gesture signifying struggle, peace, strength, and hope. Arms spread wide like the rebel in Goya's *El tres de Mayo*, kneeling before Napoleon's firing squad, or like Miyazaki's Nausicaä before the soldiers of the royal house of Pejite. An image of strength, but also of abandonment and vulnerability. This idea appeals to me: you can't expect to go forward if you're not prepared to expose yourself to chance, risk, even danger. The iconic image of the *pasionaria*, the hopelessly passionate, vulnerable artist, which occurs also in Baudoin's magnificent *Salade Niçoise*—I need it to watch over me, I need to confront it myself every day, face to face, in the person of Catherine.

Today, a day like every other, filled with appointments, discussions, meetings. This morning, a long telephone conversation with the chairman of my board of directors, the CEO of an insurance company who is deeply committed to enlisting the private sector in the service of scientific research. And this afternoon, a photo session to illustrate an interview I've given to a popular science magazine. Nothing very disagreeable about any of this. It's been more than six months since my life took a fascinating turn: new people to meet, new things to learn, new ideas to talk about.

The photographer was unpacking his equipment in my office, setting up the tripod and reflector, when the telephone rang. Absentmindedly, I picked up the receiver.

"Allô, oui."

"Hello, is this Cédric Villani?"

"Yes."

"This is Lázló Lovász calling from Budapest."

For a moment my heart stopped. Lovász is president of the International Mathematical Union—and, by virtue of his office, chair of the Fields Medal Committee. This, by the way, is all I know about the committee; apart from him, I haven't the faintest idea who's on it.

"Hello, Professor Lovász, how are you doing?"

"Good, I'm fine. I have news—good news for you."

"Oh, really?"

It was just like in a movie . . . I knew that these were the same words that Wendelin Werner had heard four years ago. But could it really be, so early in the year?

"Yes, I'm glad to tell you that you have won a Fields Medal."

"Oh, this is unbelievable! This is one of the most beautiful days in my life. What should I say?"

"I think you should just be glad and accept it."

Ever since Grigori Perelman refused the Fields Medal, the committee can't help but be uneasy: What if someone else should refuse it? But I'm hardly on Perelman's level, and I have no qualms about saying yes.

Lovász hastened to add that the laureates were being notified of the committee's decision earlier than usual to ensure that the formal announcement will come from the IMU, and not through a leak.

"It's very important that you keep it perfectly secret," Lovász went on. "You can tell your family, but that is all. None of your colleagues should know."

I shall therefore have to keep quiet for . . . six months. That's a hell of a long time! In six months (and three days) the news will be broadcast on television and radio throughout the entire world. Until then I must do everything in my power to protect this highly confidential information. I'll have to prepare myself psychologically for a marathon, not a sprint.

In the meantime, speculation about the medal winners will be rampant. But my lips shall remain sealed. As my colleague from Lyon, Michelle Schatzman, once observed, "Those who know do not speak, those who speak do not know."

Before Lovász's call, I put my chances of winning the medal at 40%. Now they've suddenly shot up to 99%! But still not to 100%—the possibility that it may have been a hoax can't be ruled out entirely. Landau himself once played a trick with a friend on a fellow physicist whom they despised. They sent him a phony telegram from the Royal Swedish Academy: "Congratulations, you've won the Nobel Prize, etc." The bastards . . .

So don't get too excited yet, Cédric. How do you know that was really Lovász on the other end? Wait until a letter arrives, formally confirming the award, before you break out the champagne!

Ah, the secret, yes—but what about the photographer in my office!?

Apparently nothing registered, he must not understand English. Let's hope not. Our session finally got under way. A picture of me posing with the mathematical physics trophy I brought home from Prague, another one of me in front of the Institute . . .

"That's good, I think I've got what we need. One thing I wanted to ask you—in the article it says that you might win a prize or something?"

"You mean the Fields Medal? That was just speculation on the interviewer's part. The announcement won't be made for quite a while, the congress doesn't take place until August."

"I see. Think you'll win?"

"Ohhh, I don't know, it's awfully hard to predict . . . no one can really say!"

•

In the years after 1918, harmony needed to be restored among the war-ravaged nations of Europe, where the Treaty of Versailles weighed heavily on Germany and the other Central Powers. What was true for society was also true for science: institutions had to be rebuilt.

In France, the mathematician and politician Émile Borel drew up plans for the Institut Henri Poincaré. In Canada, the mathematician John Charles Fields, an influential member of the recently founded International Mathematical Union, had the idea of creating an award that would serve both to recognize important work, as the Nobel Prize did, and to encourage talented younger mathematicians. It was to be embodied in the form of a medal and accompanied by a modest sum of money.

Fields donated the necessary funds, commissioned the medal's reliefs from a Canadian sculptor, and composed its inscriptions in Latin, a common language chosen to reflect the universality of mathematics.

On the obverse of the medal, Archimedes is shown in right-facing profile together with the inscription TRANSIRE SUUM PECTUS MUNDOQUE POTIRI (Rise above oneself and grasp the world).

On the reverse, laurels frame an illustration of a theorem by Archimedes on the calculation of volumes of spheres and cylinders, with the inscription CON-GREGATI EX TOTO ORBE MATHEMATICI OB SCRIPTA INSIGNIA TRIBUERE (Mathematicians gathered from all over the world have paid tribute to a remarkable work).

On the rim, the name of the laureate and the year of the award.

And the medal itself: solid gold.

Fields did not want the prize to be named after anyone, but upon his death in 1932 it was obvious to all that it should be called the Fields Medal. It was awarded for the first time four years later, in 1936, and then every four years from 1950 onward at the International Congress of Mathematicians, the grand meeting of the mathematical world. Its location changes from one occasion to the next, with as many as five thousand men and women taking part.

In keeping with Fields's wish that the prize should serve to stimulate future achievement, it is awarded to mathematicians under the age of forty. In 2006, the age-counting rule was clarified: eligible candidates must not yet be forty years old on the first day of January of the year in which the congress takes place. The number of laureates may vary between two and four, as the jury appointed by the Executive Committee of the IMU sees fit to decide.

A strict embargo on the announcement of the jury's decision, combined with careful media coordination, assures Fields Medal winners of unrivaled publicity within the mathematical community and even beyond. The medals are usually presented by the head of state in the country where the congress is held. From there the news spreads throughout the world at once.

FORTY-ONE

In the world of Paris rapid transit, each of the RER lines is remarkable in its own way. In the case of the RER B, the line I take to go to work, it would not be an exaggeration to say that it breaks down every day, and that most days it is packed until midnight or one in the morning. To be fair, it also has its virtues: it assures its passengers of regular physical exercise by making them change trains frequently, and it improves their mental agility by keeping them in suspense as to exactly when a train will reach its destination and where it will stop along the way.

But this morning, on my way home from a conference in Cairo, it is very, very early and the train is nearly empty.

The outbound flight, in the company of the cutest girl you could ever hope to meet, couldn't have been more delightful. We watched a film together on my computer, sharing earphones like brother and sister (always fly economy class, by the way, the girls are statistically cuter).

The return flight was less glamorous in every respect. Not least because I landed at Charles de Gaulle Airport after 10 p.m. (never book a flight that gets into CDG after 10 p.m., you're just asking for trouble). Too late to take the RER into Paris. And since I didn't feel like taking a taxi, almost on principle, I had to wait for the shuttle. . . . The first one was full even before it reached my stop, the second one too; as for the third one, well, maybe I could have just barely squeezed in—if I had been prepared, as some of the other people waiting in line with me were, to disregard the driver's instructions and elbow my way on. Finally got into Paris at two in the morning. By chance my old apartment in the city was empty, so I was able to get a couple hours' sleep before catching the first RER B back to the southern suburbs.

Going through my email offline, as always. Tons of messages . . . but ever since Lovász's phone call in February, and then the official letter that came by regular post a few days later, it's as though a great weight is being lifted from my shoulders. Not all at once—it will be a while longer before the sense of *urgency* leaves me—but gradually. Three and a half months from now, I'll have to deal with another kind of pressure. So I've got to try to savor this wonderful feeling of relaxation while there's still time.

One of the messages informs me that I'm the only person being considered for position no. 1928, a research appointment at the Université de Lyon-I. Good news. In any case 1928 is my lucky number: it's the year that the Institut Poincaré was founded! A transfer to Lyon-I would allow me to stay in touch with friends and colleagues in Lyon without preventing ENS-Lyon from hiring a full-time professor; the number of teaching positions there is fixed just now, they'd be in a real bind if I were to stay on.

A beggar, an older woman, is trying her luck with the few passengers on the train. She approaches me, speaking in a hoarse voice.

"Coming back from vacation with this big bag?"

"Vacation? Oh no! My last vacation was at Christmas . . . and the next one won't be anytime soon."

"Where are you coming from?"

"I was in Cairo, in Egypt, on business."

"Good for you! What is it you do?"

"I do mathematics."

"Ah, good for you. All right, so long. And good luck with the rest of your studies!"

I can't help but smile. It makes me really happy to know that people still take me for a student. But after all, she's right, I still *am* a student—maybe for as long as I live. . . .

•

Today I was in the air and "entertained" myself by taking five minutes to try to feel all the electrical, electronic, electromagnetic, aerodynamic, and mechanical phenomena that are exerted on you in and around an airplane. All these small distinct phenomena that together make up a functioning whole! It's fascinating to be aware of our surroundings . . . fascinating!

Unfortunately, at the controls of an airplane one seldom has more than 5 min. for thinking about this sort of thing.

Best wishes.

[From an email sent to me on September 9, 2010, by a complete stranger]

FORTY-TWO

Église de Saint-Louis-en-l'Île, Paris
June 9, 2010

Just a little too abruptly, I push back the censer that has been swung toward me. Black suit and black ascot as a sign of mourning; green spider pinned to my lapel as a sign of hope. I go up to the casket, touch it, bow in respect. A few inches away lies the body of Paul Malliavin, one of the foremost experts on probability theory during the second half of the twentieth century. Inventor of the famous Malliavin calculus, he did more than anyone else to unite probability with geometry and analysis. My own work on optimal transport follows in the same tradition. "In Malliavin there is Villani," as I like to remind myself from time to time.

Malliavin was a complex and fascinating person, a conservative and an iconoclast both, blessed with an exceptional mind. He took an interest in me from the beginning of my career, encouraged me, set me on my way. He also entrusted me with real responsibility as a member of the editorial board of his beloved *Journal of Functional Analysis,* which he cofounded with two American mathematicians, Ralph Phillips and Irving Segal, in 1966.

Despite the forty-eight years that separated us in age, we became friends. Our mathematical tastes were similar, and I like to think that my admiration for him was reciprocated. We never went beyond *cher ami* in addressing each other, but this was not a mere form of politeness: the sentiment was most sincerely felt.

A few years ago we both took part in a conference in Tunisia. Malliavin was already seventy-eight, but he was still so active! At the end, it was my job to summarize the conclusions of the meeting, and in a few words I tried to acknowledge his phenomenal impact on analysis and probability; I don't know if I called him a living legend, but that was the idea. Malliavin seemed a bit taken aback by my tribute. Later he took me aside and said very quietly, in that deadpan way of his, "You know, the Legend is a little tired."

Tired or not, Paul Malliavin went out swinging—doing mathematics right up until the last minute, as his son-in-law reported. He died the same day as Vladimir Arnold, another mathematical giant of the twentieth century, albeit a giant of a completely different style.

We will have to carry on without you. *Vous pouvez compter sur moi, cher ami.* The *Journal of Functional Analysis* is in good hands.

And . . . I would have been so proud to tell you about this secret phone call I received in February, I know that you would have been delighted. . . .

The ceremony's over—now to rush back to the IHP for the end of the big conference we've organized with the Clay Mathematics

Institute to celebrate Grigori Perelman's solution of the Poincaré conjecture. I absolutely *have* to get there before the last speaker finishes in order to say a few closing words. No choice but to run through the streets as fast as I can, from the Île Saint-Louis to the heart of the fifth arrondissement. If only "Monsieur Paul" could see me now, red-faced, sweating through my suit and puffing like a locomotive, he would permit himself a small smile. . . .

Wait, did I remember to bow before the casket the way you're supposed to? Well, it doesn't matter, the gesture was heartfelt. That's what counts.

•

At the turn of the twentieth century, Henri Poincaré developed an entirely new field of mathematics, differential topology, which seeks to classify the shapes of our physical environment by considering their behavior under deformation.

Deforming a ring (for example, a bagel or a doughnut) yields a cup, but never a sphere: the cup has a hole, bounded by a handle; the sphere does not. Generally speaking, in order to understand surfaces (the shapes on which a point is picked out within a small area by two coordinates, such as longitude and latitude), you have only to count the number of handles—the number of holes.

As it happens, we live in a world of three spatial dimensions. To classify such objects, is it really enough to count the number of holes? This is the question Poincaré posed in 1904 by way of conclusion to a series of six articles, notable as much for their untidiness as for their sheer genius, that gave birth to the infant discipline of topology. Poincaré asked, in effect, whether all shapes of dimension 3 that are bounded (finite universes, as they might be called) and have no holes are equivalent. One such shape naturally suggested itself for study: the 3-sphere, a sphere with three coordinates embedded in a space of dimension 4. Poincaré formulated his conjecture thus: A smooth manifold of dimension 3, compact and without boundary, simply connected, is diffeomorphic to the 3-sphere.

This statement seems perfectly plausible. But is it true? Poincaré answered with these tantalizing words, almost as memorable as Fermat's famous complaint about the margin that was too narrow: "But this question would take us too far afield."

Ages and ages passed. . . .

Poincaré's conjecture became the most celebrated enigma in all of geometry, nourishing research for the whole of the twentieth century. In the interval no fewer than three Fields Medals were awarded for steps toward a solution.

A decisive stage was reached when the American mathematician William Thurston turned his mind to the problem in the early 1980s. Thurston was a visionary. His extraordinary geometric intuition enabled him to see all the shapes—all the "possible universes"—of dimension 3. What he proposed was a sort of zoological, or taxonomic, classification of three-dimensional shapes. Its magnificence dazzled even skeptics: those who still had their doubts about Poincaré nonetheless bowed down before a conception that was so beautiful it had to be true. What is more, proving the so-called Thurston program (or geometrization conjecture) would imply the truth of Poincaré's conjecture. Thurston himself was able to explore only a part of the vast landscape he had surveyed.

It came as no surprise when Poincaré's conjecture was chosen by the Clay Mathematics Institute in 2000 as one of seven problems for whose solution it was offering a prize of $1 million apiece. At the time it was widely believed that there was a good chance the famous problem would remain unsolved for another century!

Only two years later, however, in 2002, Grigori Perelman dumbfounded everyone by announcing a solution of Poincaré's conjecture. A solution on which he had worked alone, in secret, for seven years!!

Grigori Perelman

Born in 1966 in Leningrad (alias Saint Petersburg), Perelman had inherited a passion for mathematics from his mother, a college-level teacher. But he was decisively influenced at an early age by two things: the exceptional Russian school of mathematics, led by Andrei Kolmogorov; and his local math club, whose intensely devoted coaches prepared him as a high school student for the International Olympiad. He went on to study with three of the best geometers of the century—Alexandrov, Burago, Gromov—and quickly became a leading expert on the theory of singular spaces with positive curvature. In 1994, his proof of the so-called soul conjecture brought him wide recognition. Seemingly destined for a brilliant career, Perelman then—disappeared!

From 1995 on there was no sign of him. Far from giving up mathematics, however, he had begun thinking about Richard Hamilton's seminal work on the Ricci flow, a formula that makes it possible to continually deform geometric objects by stretching their curvature, in the same way that the heat equation spreads out temperature. Hamilton had hoped to be able to use his equation to prove Poincaré's conjecture, but seemingly insuperable technical problems hampered his progress for many years. The way forward was blocked. . . .

Until the famous email message that Perelman sent on November 12, 2002, to a select group of mathematicians in the United States. A message of only a few lines, calling their attention to the preprint of a paper he had just made publicly available on the Internet, which contained, as he put it, a "sketch of an eclectic proof" of Poincaré's conjecture—and, indeed, of a large part of the Thurston program.

Borrowing an idea from theoretical physics, Perelman showed that a certain quantity (which he called "entropy" because it resembles Boltzmann's concept of the same name) decreases when its geometry is deformed by means of the Ricci flow. Thanks to this original insight, whose ramifications have probably not yet been fully appreciated, Perelman was able to prove that the Ricci flow can go on and on without ever "blowing up," that is, without producing an uncontrollably violent singularity. Or rather: before any such singularity develops, its occurrence can be foreseen and counteracted.

Perelman then came to the United States to give a series of lectures about his work, and impressed everyone who heard them with his mastery of the problem.

He was angered, however, by the inaccuracy of media accounts and the behavior of American academic institutions, and annoyed by how long it took his fellow mathematicians to grasp the arguments he had presented. He went back to Saint Petersburg, leaving it to others to check the soundness of his reasoning. In the event, it took almost four years for three teams of researchers to verify Perelman's proof and make it complete down to the last detail!

The enormity of what was at stake, as well as Perelman's retreat once more from professional life, placed the mathematical community in an unprecedented situation. Yet despite the tensions aroused by controversy over the proper share of credit due to Perelman, it eventually became clear that he had indeed proved Thurston's grand geometrization conjecture and, along with it, Poincaré's conjecture. This feat has no equivalent in recent memory, unless perhaps Andrew Wiles's proof of Fermat's last theorem fifteen years ago.

Perelman was showered with honors: the Fields Medal in 2006, followed at once by recognition of his achievement in the journal Science *as the "Breakthrough of the Year" (an accolade almost never conferred on a mathematician), and, four years later, in 2010, the Clay Millennium Prize, making him the first recipient of the most richly endowed prize in mathematics. Perelman had no need of awards and turned them down, one after another.*

Editorial writers everywhere fell over themselves in their eagerness to explain his refusal of $1 million, tirelessly harping on the theme of the mad mathematician. All of them missed the point: what was exceptional in Perelman's case was neither his refusal of money and honors nor his undoubted personal eccentricity—many other such examples can be found in the annals of mankind—but instead the extraordinary force of character and intellectual brilliance required to persevere, through seven long years of courageous and solitary labor, and finally to penetrate the outstanding mathematical enigma of the twentieth century.

In June 2010, the Clay Mathematics Institute and the Institut Henri Poincaré jointly organized a conference in Paris in honor of this remarkable feat. Fifteen months later they announced that the money Perelman had refused would be used to create a very special position at the IHP. The Poincaré Chair, as it was named, is to be held by a succession of extremely promising young mathematicians under ideal conditions, unencumbered by any obligation to give lectures or to reside in Paris for the term of the appointment. The perfect opportunity, in other words, for them to realize their full potential—just as Perelman himself had been allowed to do as a guest of the Miller Institute at Berkeley some fifteen years earlier.

FORTY-THREE

My name rings out in the immense hall, and my portrait—by the photographer Pierre Maraval, with carmine red cravat and mauve-tinted white spider—is displayed on a gigantic screen. I didn't sleep last night, and yet I can't remember ever having felt so alert in all my life. This is the most important moment of my professional career, the moment every mathematician dreams of without daring to admit it. A more or less anonymous young man, number 333 of the one thousand scientists photographed by Maraval, now begins to emerge from obscurity. . . .

I get up and make my way to the stage while the citation is read out: *A Fields Medal is awarded to Cédric Villani for his proofs of nonlinear Landau damping and convergence to equilibrium for the Boltzmann equation.*

I mount the steps, not too slowly, not too quickly, and approach the president of India at the center of the stage. Petite though she is, Shrimati Pratibha Pati radiates an aura of power visible in the bearing of her entourage. I stop just in front of her: she bows slightly and I bow in return, much too deeply. *Namaste.*

She hands me the medal and I hold it up for all to see. My torso is unnaturally contorted: making a full ninety-degree turn to face the audience would force the chief of state to look at me from the side, so I do my best to split the difference, forty-five degrees for each of them.

Three thousand people applauding in this vast meeting hall

attached to a luxury hotel, site of the 2010 International Congress of Mathematicians. How many were there, eighteen years ago, applauding my welcoming address at the bicentennial ball of the École Normale Supérieure? A thousand, maybe? *Those were the days, my friend.* . . . My father was so disappointed he couldn't film the ceremony; he'd left his camera in a room somewhere and the organizers wouldn't let him go back and get it. The ball in Paris was quite an affair, but small potatoes in comparison with this. There's a whole army of photographers and cameramen besieging the stage—it's like the Cannes Film Festival!

Clutching the medal to my breast, I bow once more to the president, take three steps back, pivot, and walk toward the wings at a measured pace, almost exactly as we had been instructed to do at the rehearsal yesterday evening. . . .

Not bad. In any case I survived the ordeal better than Elon Lindenstrauss, the first of the four of us to be honored, who made a shambles of all the fine points of etiquette. Completely out of it. As he shuffled off to the side, Stas Smirnov whispered in my ear, "We can't possibly do worse."

The four of us, standing together, were immortalized by the official photographers at the end of the ceremony. After that it's all a blur. All I know for sure is that many more pictures were taken. Back down on the floor, we held up our medals to the digital swarm—video cameras, camera phones, capturing and registering devices of every imaginable kind. And then there was a press conference. . . .

Clément is here, of course, beaming. To think it wasn't even ten years ago that he first set foot in my office at ENS-Lyon, looking for a thesis topic. . . . A stroke of luck for him, a stroke of luck for me.

No computers or cell phones were allowed in the main hall. Soon I discovered three hundred messages in my mailbox. Many more were on the way. Congratulations from colleagues, old friends whom I haven't seen for ten, twenty, or thirty years, perfect strangers, classmates from way back in elementary school . . . Some of the messages were very moving. One informed me of the death, several years ago, of a childhood friend. Life, as we all know, is filled with joys and sorrows, inextricably entangled.

And via the press, an official message of congratulations from the president of France. As expected, Ngô also won a medal. It took me a while to fully appreciate how proud our fellow citizens are of this double victory—to say nothing of the fact that Yves Meyer was awarded the prestigious Gauss Prize for lifetime achievement! People back home are now realizing that for more than three centuries France has been at the forefront of international mathematical research. As of this evening, our country has produced no fewer than eleven of the fifty-two Fields Medal winners!

I fought my way through the crowd and went up to my hotel room. A dull, uninteresting room with nothing of India about it—I might just as well be in Tierra del Fuego! But I'm here to discharge my duty.

For four straight hours I answered calls from journalists, switching back and forth between fixed and mobile phones. Not a moment's rest. No sooner had one call ended than I checked my voicemail and found new messages. Personal questions, scientific questions, institutional questions. And questions that basically asked the same thing, over and over again: *How does it feel to win this award?*

Finally I took the elevator back downstairs, looking a little pale, feeling a little hungry—but there are worse things to be endured, after all. I settled happily for a cup of masala chai and then plunged back into the crowd. Throngs of young people, most of them Indian, clamored for my attention. Dazed from signing hundreds of autographs and posing for at least as many pictures, I somehow made it through the end of the day. . . .

Unlike the other laureates, I came here alone. I thought it would be best if Claire and the children were to stay home in France, far from the noise and the tumult of the ICM. And I was right! In the meantime I had faithfully obeyed my orders and told no one about the medal except my wife. Not even my parents, they learned of it only when journalists phoned them for their reaction!

And . . . Catherine Ribeiro sent a superb bouquet of roses to my home!

Never for a moment could I have imagined that while I was basking in the limelight in Hyderabad, hordes of shutterbugs snapping away, Michelle Schatzman lay dying back in Lyon. Daughter of the great French astrophysicist Évry Schatzman, Michelle was one of the most original mathematicians it has been my good fortune to know, eager to accept whatever challenge to her abilities the classroom could devise while at the same time exploring frontiers of research where no one else would dare to go, especially the one that lies between algebraic geometry and numerical analysis. Indeed, this was the title she gave to a manifesto she dashed off one day, as though it were something perfectly obvious—*Frontières*. Michelle was my friend from the moment I came to Lyon in 2000; we went to seminars together, and more than once plotted together to attract a first-rate mathematician to the faculty at the Université de Lyon.

Michelle Schatzman

Michelle never shrank from speaking her mind, even if it meant putting her foot in it, as she not infrequently did. Her scathingly black humor was legendary. For more than five years she battled an incurable cancer, undergoing both chemotherapy and surgeries. With a glint in her eye she told us how good life was now that she didn't have to spend money on shampoo. A few months ago we celebrated her sixtieth birthday with a workshop in Lyon. The speakers came from near and far. Among them the polymorphous Uriel Frisch, a world-renowned physicist who had been a student of Michelle's father; and myself, the spiritual son of one of Frisch's spiritual sons, Yann Brenier, who was also there. Michelle brilliantly suggested a connection between my talk on Landau damping and the "tygers" that Uriel had discussed. Pure elegance!

But then suddenly a few weeks ago her condition began to deteriorate. As proud and forthright in sickness as she had been in health, Michelle refused morphine in order to go on thinking clearly right until the end. She had been impatiently awaiting the results of the Fields Medal competition. On her deathbed she learned that I had won; a few hours later she passed away. Life, as we all know, is filled with joys and sorrows, inextricably entangled. . . .

•

On August 19, 2010, the Hyderabad International Convention Centre in India contained within its walls the greatest concentration of mathematicians in the world. They came from every continent, bringing with them their many and varied talents: experts in analysis, algebra, probability, statistics, partial differential equations, algebraic geometry and geometric algebra, hard logic and soft logic, metric geometry and ultrametric geometry, harmonic analysis and harmonious analysis, the probabilistic theory of numbers and numina; discoverers of models and supermodels, surveyors of macroeconomies and microeconomies, designers of super-computers and genetic algorithms, processors of images and developers of Banach spaces. Mathematics of the summer, of the fall, of the winter, of the spring: a myriad of specialities that transform their masters into the Great God, Shiva, the god with a thousand mathematical arms.

One after the other, the Fields medalists, together with the winners of the Gauss, Nevanlinna, and Chern prizes, were offered up in sacrifice to Shiva. The high priestess, the president of India, presented the seven terrified laureates to the ecstatic crowd.

This was the beginning of the great festival of the International Congress of Mathematicians, which over a period of ten days or so witnessed a nonstop succession of talks and discussions, cocktail parties and receptions, interviews, photo sessions, and evenings filled with dancing and laughter. Revelers swanning about from one event to the next in luxury limos and romantic rickshaws, everywhere celebrating the unity and diversity of mathematics, its ever-shifting shapes and forms; everywhere overcome with joy at what has so far been accomplished and wonder before all that yet remains to be discovered; everywhere, day and night, dreaming of the unknown.

Once the festival is over, the celebrants will go back to their universities and research centers, to their campuses, business parks, and home offices, and resume once more, each in his or her own way, the great adventure of mathematical exploration. Armed not only with their logical abilities and their appetite for hard work, but also with their imagination and their passion, they will be joined together once more in a common desire to push back the frontiers of human knowledge.

And already they are thinking of the next congress, four

years from now, in the lair of the Korean tiger. What will they talk about? Who will be the next sacrificial victims?

Four years from now, thousands of mathematicians will gather in Seoul to pay their respects to the venerable tiger. They will explore its sinuous geometry, axiomatize its implacable symmetry, test its turbulent stochasticity, analyze the reaction–diffusion to which it owes its stripes, perform differential surgery on its powerful paws, measure the curvature of its sharp claws, release it from the potential wells of quantum mechanics, and get high smoking ethereal theories that turn its whiskers into vibrating strings. For a few days the tiger will be a mathematician, from the end of its tail all the way to the tip of its nose.

[My contribution to the Korean edition of
Les déchiffreurs: Voyage en mathématiques]

•

TYGER PHENOMENON FOR THE GALERKIN-TRUNCATED BURGERS AND EULER EQUATIONS

It is shown that the solutions of inviscid hydrodynamical equations with suppression of all spatial Fourier modes having wave numbers in excess of a threshold K_g exhibit unexpected features. The study is carried out for both the one-dimensional Burgers equation and the two-dimensional incompressible Euler equation. At large K_g, for smooth initial conditions, the first symptom of truncation, a localized short-wavelength oscillation which we call a "tyger," is caused by a resonant interaction between fluid particle motion and truncation waves generated by small-scale features (shocks, layers with strong vorticity gradients, etc.). These tygers appear when complex-space singularities come within one Galerkin wavelength $\lambda_g = 2\pi/K_g$ from the real domain and typically arise far away from preexisting small-scale structures at locations whose velocities match that of such structures. Tygers are weak and strongly localized at first—in the Burgers case at the time of the appearance of the first shock their amplitudes and widths are proportional to $K_g^{-2/3}$ and $K_g^{-1/3}$ respectively—but grow and eventually invade the whole flow. They are thus the first manifestations of the thermalization predicted by T. D. Lee [in 1952]. The sudden dissipative anomaly—the presence of a finite dissipation in the limit of vanishing viscosity after a finite time—, which is well known for the Burgers equation and sometimes conjectured for the 3D Euler equation, has as counterpart in the truncated case: the ability of tygers to store a finite amount of energy in the limit $K_g \to \infty$. This leads to Reynolds stresses acting on scales larger than the Galerkin wavelength and thus prevents the flow from converging to the inviscid-limit solution. There are indications that it may be possible to purge the tygers and thereby to recover the correct inviscid-limit behaviour.

[Abstract of article published by Samriddhi Sankar Ray,
Uriel Frisch, Sergei Nazarenko, and
Takeshi Matsumoto in 2011]

•

THE TYGER

Tyger! Tyger! burning bright
In the forests of the night,
What immortal hand or eye
Could frame thy fearful symmetry?

In what distant deeps or skies
Burnt the fire of thine eyes?
On what wings dare he aspire?
What the hand, dare seize the fire?

And what shoulder, & what art,
Could twist the sinews of thy heart?
And when thy heart began to beat,
What dread hand? & what dread feet?

What the hammer? what the chain?
In what furnace was thy brain?
What the anvil? what dread grasp
Dare its deadly terrors clasp?

When the stars threw down their spears,
And water'd heaven with their tears,
Did he smile his work to see?
Did he who made the Lamb make thee?

Tyger! Tyger! burning bright
In the forests of the night,
What immortal hand or eye
Dare frame thy fearful symmetry?

[From William Blake, *Songs of Experience*, 1794]

FORTY-FOUR

Autumn. Everything's gold, red, and black: golden leaves, red leaves—and shiny black ravens, like the ones in Tom Waits's November song.

I get off at my station on the dear old RER B line and disappear into the night.

The last three months have been so intense!

The autographs.

The newspaper articles.

The radio interviews.

The television shows.

The documentaries.

My appearance with Franck Dubosc, whom I met for the first time doing a live show on Canal+ . . . Some critics reproached me for taking part in a "farce," but where's the harm in talking to people who think they have no interest whatsoever in mathematics? The next day perfect strangers stopped me in the street: "Hey, I saw you on TV last night!"

And the meetings with politicians, with artists, with students, with industrialists, with business executives, with revolutionaries, with parliamentarians, with senior civil servants, with the President of the Republic . . .

Questions that run together into one long question: *How did you get interested in math why are the French so good at math did the Fields Medal change your life what keeps you interested now that you've received the highest honor are you a genius what is the meaning of your spider. . . . ?*

Ngô has gone back to the United States, leaving me to face the onslaught alone. I don't mind. It's fascinating to get a glimpse of these different worlds—behind the television cameras, inside the newsroom of a big daily paper. I've seen first-hand how an interview frequently takes on a life of its own, separate from what the person being interviewed actually says; how an abstract media personality named Cédricvillani comes to be created, someone who's not really me and whom I can't really control.

All the while continuing to do the job I'm paid to do as director of the IHP. The same day I appeared with Dubosc on Canal+, I'd already done an interview with RTL, attended a meeting at the Hôtel de Ville on university housing, had a long conversation with the chairman of my board of directors, and recorded a show for *Des Mots de Minuit*.

A lot of my time has been taken up guiding a joint effort to obtain major funding through the government's Investments for the Future program (aka "The Big Loan"), a complicated business that requires coordinating the interests of the four national and international institutes of mathematics in France: the Institut Henri Poincaré (IHP), the Institut des Hautes Études Scientifiques (IHÉS) in Bures-sur-Yvette outside Paris, the Centre International de Rencontres Mathématiques (CIRM) in Luminy, and the Centre International de Mathématiques Pures et Appliquées (CIMPA) in Nice.

The IHÉS is the French version of the Institute for Advanced Study in Princeton: a magnificent rural retreat where the autumn air crackles with the sound of chestnuts falling to the ground, where the fantastically brilliant Grothendieck produced the better part of his incomparable work, and where talented young people can accelerate the pace of their research through contact with some of the best mathematicians in the world. CIRM, with its weeklong conferences, is the French counterpart to the institute in Oberwolfach, except that here the austere beauty of the Black Forest has been replaced by the deep and rocky inlets around Marseille, no less splendid in their way. CIMPA, for its part, is a thoroughly international organization devoted to supporting the study and use of mathematics, mainly in developing countries but anywhere, really, that its assistance is both needed and welcome.

The governing bodies of these four institutions are very different. Getting them to agree to collaborate on this project took hours and hours of negotiation. After a year at the helm of the IHP, with a few bureaucratic skirmishes under my belt, I felt ready to step forward and take responsibility for coordinating the joint initiative. Our group is to be called CARMIN, for Centre d'Accueil et de Rencontres Mathématiques Internationales (Reception Center for International Mathematics Meetings).

In my spare time I composed and delivered two public lectures on mathematics, part of an ongoing series, and, appropriately enough, wrote a long paper on the subject of time for a theoretical physics seminar. On top of all this I had to take on extra administrative duties to help the Institute get through a rough patch when, by a sort of curse, several staff members fell sick at the same time. Fortunately for me, everyone else pitched in and worked twice as hard as well!

These three months have worn me out. There were times I had to plan my sleeping schedule several days in advance. *Hasta que el cuerpo aguante!*

Thinking back on this exhausting autumn as I walk home along the dirt path from the RER station . . . now I come to the **dark** part of my journey.

To my left, a forest, with foraging foxes and deer; to my right, a field, cows peaceably slumbering; in front of me, for the next three hundred yards, complete darkness. No public lighting, not the least speck of luminous pollution.

Nothing is more precious than an unlit path! When the moon is hidden, you can't see even ten feet ahead. You walk a bit faster, your heart beats a little more quickly, your senses are in a heightened state of alert. The slightest noise makes your ears prick up. You tell yourself that the way home seems longer than usual. You imagine a robber lying in wait. You try not to run.

This gloomy tunnel is a little like the one you pass through when you begin work on a new mathematical problem. *A mathematician is a blind man in a dark room looking for a black cat that isn't there . . .* (as Darwin may or may not have said). Total obscurity. Bilbo in Gollum's tunnel.

A mathematician's first steps into unknown territory constitute the first phase of a familiar cycle.

After the darkness comes a faint, faint glimmer of light, just enough to make you think that something is there, almost within reach, waiting to be discovered. . . . Then, after the faint, faint glimmer, if all goes well, you unravel the thread—and suddenly it's broad daylight! You're full of confidence, you want to tell anyone who will listen about what you've found.

And then, after day has broken, after the sun has climbed high into the sky, a phase of depression inevitably follows. You lose all faith in the importance of what you've achieved. *Any idiot could have done what you've done, go find yourself a more worthwhile problem and make something of your life.* Thus the cycle of mathematical research . . .

For the moment I'm making my way through the darkness, literally, recollecting the events of a day of great emotion. Ngô, Meyer, and I met with the president of the National Assembly, in whom we recognized a comrade in arms the moment we became aware of his scientific background; then, just before question time got under way, we were acclaimed by the whole Assembly. Earlier, in the library, I had been allowed to see an indescribable treasure, a massive piece of furniture specially designed to hold the works composed by the scientists, engineers, and scholars who accompanied Napoleon on the expedition to Egypt. Works by Monge, Fourier, and so many others whose findings revolutionized our understanding of natural history, physical geography, archaeology, ancient technology, you name it. The beauty of the illustrations, drawn by hand, using improvised tools; the majesty of these extraordinary oversized volumes, which normally only highly qualified conservators are allowed to handle—all of this deeply moved me, imbued me with a warm inner glow.

And yet, in the back of my mind, a tiny seed of doubt has grown little by little over the past few months into a nagging worry. Still no word from *Acta*! Still no word from the referees! Impartial review by experts whose anonymity is carefully protected: this, and only this, can confirm or disconfirm our results.

After all of the honors I have received, what will I say if our findings turn out to be wrong? The Fields committee must have taken care to check the explanation of Landau damping, knowing full well what was at stake—but as usual I have no idea what is really going on. What if, after long and patient scrutiny, an error were to be detected, or if a favorable report were to be contradicted by a second, still more thorough round of review?

Cédric, you're a father—ritual suicide is not an option.

Seriously, though, everything is going to work out just fine. And besides, I'm almost through the tunnel of darkness. There it is, all the way at the end, a faint, faint, flickering glimmer—the light of the digicode panel. I punch in my security number on the keypad at the front gate. Made it!

It feels so good every time I make it home safely through this black forest, there's nothing like it! I push open the heavy wrought-iron gate, cross the courtyard, unlock the front door, turn on the lights, go upstairs to my office, plug in my laptop, and download my emails. What, only eighty-eight new messages in the past twelve hours?! Slow day . . .

But there, buried in the middle of the list, one name immediately attracts my notice: *Acta Mathematica*! I feverishly open the message from Johannes Sjöstrand, the editor handling our paper.

The news about your paper are good.

He should have written "is," of course: *news*, like *mathematics*, is singular despite the final *s*. But who cares? I don't need to read any further. I forward the message at once to Clément, adding two words: *Gooood news.*

Our theorem has truly been born at last.

•

Theorem (Mouhot and Villani, 2009)

Let $d \geq 1$ be an integer, and let $W : \mathbb{T}^d \to \mathbb{R}$ be a locally integrable, even periodic function whose Fourier transform satisfies $|\hat{W}(k)| = O(1/|k|^2)$.

Let $f^0 = f^0(v)$ be an analytic distribution $\mathbb{R}^d \to \mathbb{R}_+$, such that

$$\sum_{n \geq 0} \frac{\lambda_0^n}{n!} \| \nabla_v^n f^0 \|_{L^1(dv)} < +\infty,$$

$$\sup_{\eta \in \mathbb{R}^d} \left(|\tilde{f}^0(\eta)| e^{2\pi \lambda_0 |\eta|} \right) < +\infty$$

for some $\lambda_0 > 0$, where $\tilde{f}$ designates the Fourier transform of f.

Assume that W and f^0 satisfy the generalized Penrose linear stability condition: for all $k \in \mathbb{Z}^d \setminus \{0\}$, if we assume further that $\sigma = k/|k|$ and for all $u \in \mathbb{R}$, $f_\sigma(u) = \int_{u\sigma + \sigma^\perp} f^0(z) dz$, then for all $w \in \mathbb{R}$ such that $f'_\sigma(w) = 0$, we have

$$\hat{W}(k) \int_\mathbb{R} \frac{f'_\sigma(u)}{u - w} du < 1.$$

Now assume an initial position and velocity profile, $f_i(x, v) \geq 0$, very close to the analytic state f^0, in the sense that its Fourier transform $\tilde{f}_i$ with respect to position and velocity satisfies

$$\sup_{k \in \mathbb{Z}^d, \eta \in \mathbb{R}^d} |\tilde{f}_i(k, \eta) - \tilde{f}^0(\eta)| e^{2\pi \mu |k|} e^{2\pi \lambda |\eta|}$$

$$+ \iint |f_i(x, v) - f^0(v)| e^{2\pi \lambda |v|} dx \, dv \leq \varepsilon$$

with $\lambda, \mu > 0$, and $\varepsilon > 0$ small enough.

Then there exist analytic profiles $f_{+\infty}(v)$, $f_{-\infty}(v)$ such that the solution of the nonlinear Vlasov equation, with interaction potential W and initial datum f_i at time $t = 0$, satisfies

$$f(t,\cdot) \xrightarrow{t \to \pm\infty} f_{\pm\infty}$$

weakly; more precisely, in the sense of a pointwise, exponentially fast convergence of Fourier modes.

The convergence rate of the nonlinear equation is arbitrarily close to the convergence rate of the linearized equation, so long as $\varepsilon > 0$ is sufficiently small. Additionally, the marginals $\int f\, dv$ and $\int f\, dx$ converge exponentially fast to their equilibrium value, in all C^r spaces.

All the estimates appearing in the nonlinear statement are constructive.

Clément Mouhot

Cédric Villani

EPILOGUE

Four bottles lined up in a row on a small rickety table. My mind most pleasantly clouded by a fine red wine from the famous Villány region, I strain to follow Gábor's detailed analysis of the comparative virtues of these four white wines. Young, dry, sweet . . . I am incapable of choosing among them.

After two helpings of goulash and apple pie, the children run off to take pictures of everything in the small apartment, dominated by a giant screen in the main room. Claire helps me select a sweet, organic Tokay. Gábor's wife, Réka, serves a superb cappuccino made with deliciously creamy milk.

Gábor speaks of his native land, and of his childhood, when kids happily spent twelve hours a week doing mathematics and the Olympiad problems were announced on television every year—problems that Réka still remembers.

He speaks of his extraordinary language, a distant cousin of Finnish, from which it branched off a thousand years ago. A language that forces the listener to be constantly alert, constantly wondering whether the next word is going to utterly transform the meaning he has registered so far. Is this why Hungary produced more legendary mathematicians and physicists than any other country during the first half of the twentieth century? The land of Erdős, von Neumann, Fejér, Riesz, Teller, Wigner, Szilárd, Lax, Pólya, and all the others.

"The Jews played a vital role!" Gábor insists. "For a time our country was the least anti-Semitic in this part of the world. Jewish thinkers flocked to Hungary and dramatically increased its intellectual wealth. Then the winds shifted: they were no longer welcome, and, alas, they went away. . . ."

Gábor discovered the Gömböc, an incredible shape that Vladimir Arnold felt sure must exist: solid, homogeneous, having only one stable point of equilibrium and one unstable point. A minimal superstable shape, in other words, which always comes back to its equilibrium position no matter how you set it down on a flat surface. Like a Weeble—except the Weeble is weighted at the bottom, whereas in a Gömböc the weight is evenly distributed.

The moment I arrived in Budapest, I heard people talking about this object that Gábor Domokos had discovered. I pictured a Gömböc on display in the library of the IHP in Paris. But first of all I wanted to see it, to convince myself that such a thing really exists! A quick exchange of emails was all it took. My institute would be very honored to exhibit your marvelous object. I would be very honored if my discovery were to be recognized by your prestigious institute, I'll be at your talk tomorrow, would you be able to come to my house for lunch the day after? Sounds great, I'm eager to meet you.

"What a beautiful talk you gave yesterday at the university!" Gábor exclaimed. "What a talk! It was so beautiful, one would have sworn that Boltzmann himself was in the room. In our midst! I'm not kidding, what a beautiful talk!"

He described the scene for Claire: "The room was overheated and too small for the audience, the projector didn't arrive, your husband had to be careful he didn't trip over all the wires that were lying around, the blackboard wouldn't stay in place, it kept slipping—but it didn't bother him in the least! He talked for an hour and a half! What a pleasure!"

We raised our glasses to Boltzmann, to the brotherhood among mathematicians of all countries—and to my article on Landau damping, which, after a bit more toing and froing with the referees, was officially accepted yesterday for publication by *Acta Mathematica*.

The sweet Tokay was going down easily, Gábor went on talking. About his experience at the International Congress on Industrial and Applied Mathematics in Hamburg fifteen years ago, in 1995. A luncheon had been organized with Arnold: everyone except the guest of honor had to pay. Gábor signed up at once—even though it would cost him half of his miserably small travel budget. Came the day, he was so intimidated he didn't dare even to speak to the great man!

But the next day, quite by chance, Gábor crossed paths with his hero again. Arnold was unsuccessfully trying to fend off a pest (I solved your problem ten years ago, I really don't have time to listen to your proof) and immediately took advantage of Gábor's unexpected appearance in order to extricate himself (No, really, I'm sorry, I have an appointment with this gentleman here).

As it turned out, Arnold was curious to know more about the strangely silent guest from the day before. "I saw you at the luncheon yesterday. I know you're from Hungary and that the meal was very expensive for you, so if you've got something to say to me, now's a good time!"

Gábor described his research and Arnold told him he was going about it the wrong way. In the course of their conversation, Arnold confided his belief in the existence of a minimum stable shape: a shape having two equilibria, only one of which is stable.

These few minutes changed Gábor's life. Little did he imagine that he would spend fully twelve years seeking the famously elusive shape. Together with Réka, he collected and experimented with thousands of pebbles before concluding that it didn't exist in nature and would have to be created. Perhaps by deforming a sphere to produce a spheroid—since true spheroids are seldom found in nature.

Finally, in 2006, Gábor found the shape he was looking for with the help of his student Péter Vàrkony, whom chance had made his partner. He named it Gömböc, the Hungarian word for "spheroid."

The first Gömböc was abstract, so close to a sphere that the difference was imperceptible to the naked eye. But gradually its creators succeeded in deforming it more and more, to the point that it looked like a cross between a tennis ball and a prehistoric stone carving, while yet retaining the crucial property of having only one stable and one unstable equilibrium position!

Gábor hands me an enormous Gömböc made out of Plexiglas.

"Isn't it beautiful? Twelve years of research! When the Chinese see it they think it's a three-dimensional representation of yin and yang! I gave the very first one to Arnold for his seventieth birthday. I'll send you a fine specimen, cast in metal—and numbered 1928, the date your institute was founded."

Another glass of Tokay. The children are taking pictures of the images streaming across the giant screen. Réka is taking pictures of the children taking pictures. Gábor goes on talking. I go on listening, fascinated by his story. An eternal story, a mathematical story, of quests, of dreams—and of passion.

A Note on the Translation

The English translation of *Théorème Vivant* differs in many small ways from the second French edition, incorporating additional corrections and clarifying certain passages. Still, the temptation to be as precise as possible has been resisted. The book is meant chiefly as a work of literary imagination. Whatever else it may be, it is in no way, shape, or form a scientific treatise.

At the author's suggestion, a brief section of notes has been included as a courtesy to readers who may be unacquainted with some of the references in the text, particularly to aspects of popular culture. No attempt has been made to expand upon, much less to explain, fine points of mathematical detail, many of which will be unfamiliar even to professional mathematicians. The technical material, though not actually irrelevant, is in any case inessential to the story Cédric Villani tells in this book.

Notes

THREE

18 **the fiendishly gifted surgeon:** Black Jack is a character created by the Japanese artist and animator Tezuka Osamu (1928–89), the legendary "father of manga."

18 **would have resumed his superhuman labors:** In January 1962 Landau was nearly killed in a car accident, which left him in a coma for two months. He never completely recovered from his injuries, and died six years later.

FIVE

32 **the ones that David B. records:** The pen name of the French artist and writer Pierre-François ("David") Beauchard.

35 *Vincent Beffara isn't here:* In real life, Beffara is a mathematician at ENS-Lyon.

SIX

39 *Gott im Himmel!:* Echoing Kurt Weill's "Ballad of Marie Saunders."

39 *Cédurak go!:* A personal slogan from the author's school days at Lycée Louis-Le-Grand in Paris, adapted from the command ("Goldorak go!") issued by the space-fighter hero of a popular Japanese anime of the same name from the late 1970s.

40 **(chapter 3, exercise A.1.a):** Villani gives a corrected version of the exercise that appears in the original 1991 edition of Alinhac and Gérard's book, the one he actually consulted. A revised English edition was published by the American Mathematics Society in 2007, but the alterations made by the authors to the statement of this exercise were apparently marred by typos—hence the very minor revisions silently made here.

EIGHT

52 **Even the name of the formula's author I had forgotten:** The formula is named after the Italian priest Francesco Faà di Bruno (1825–88), beatified by

the Church on the centennial of his death. Faà di Bruno studied mathematics under Cauchy and Le Verrier in Paris.

54 *(Arbogast 1800, Faà di Bruno 1855)*: The formula appears to have been first stated by the French mathematician Louis François Antoine Arbogast (1759–1803).

NINE

60 **the project with Alessio and Ludovic:** Alessio Figalli, then at the École Polytechnique in Palaiseau (now at the University of Texas–Austin), and Ludovic Rifford, at the Université de Nice–Sophia Antipolis.

60 *On stories I really like*: A line from William Sheller's 1979 single "Oh! J'cours tout seul."

62 **The general criterion we've proposed for curvature-dimension:** Lott–Sturm–Villani spaces, as they are now called—after work done by the author in collaboration with John Lott, at Berkeley, and independently by Karl-Theodor Sturm, at the University of Bonn.

TEN

65 **Claire's turn to get hooked on it:** The manga *Death Note*, written by Tsugumi Ohba and illustrated by Takeshi Obata, was serialized between 2003 and 2006, when the first anime episodes, directed by Tetsuro Araki, were aired on Japanese television. A DVD edition followed shortly thereafter.

ELEVEN

70–71 **Fischer and Erdős were Hungarian:** Erdős was fully Hungarian by birth, Fischer partially (his biological father seems almost certainly to have been a Hungarian Jewish physicist named Paul Nemeny).

70 **Alexander Grothendieck, a living legend:** Grothendieck passed away, not long before this book went to press, on November 13, 2014.

72 **From the liner notes to *Final Report* (1999):** Excerpted from the brief history of Do Not Erase written by the historian Marshall Poe (vocals, rhythm guitar) that appears on the cover of its only recording. The other members of DNE were mathematicians: Carol Namkoong (drums, backing vocals), David Renard (lead guitar), Konstanze Rietsch (vocals), Peter Trapa (bass), and Lindi Wahl (vocals). The full text of the liner notes is available at www .math.utah.edu/~ptrapa/finalreport/linernotes.html.

TWELVE

75 **I'll take care of the little lambkins:** The pet name in French is *poutchou*, thought to be a shortened form of the expression *petit bout de chou* (literally, a

tiny cabbage) and used affectionately to mean a "little darling." It may be related to *pitchoun*, the Provençal word for "little" or "little one."

SIXTEEN

96 **There was even some Rove:** A cheese made from the milk of a rare breed of goat, originally raised in the village of this name in the south of France, near Marseille.

99 **From the mission statement of the Institut Henri Poincaré:** Drafted by its new director in September 2010.

SEVENTEEN

101 *My heart will conquer without striking a blow:* "Mon cœur vaincra sans coup férir." From Guillaume Apollinaire's poem "L'espionne," in *Calligrammes: Poèmes de la paix et de la guerre* (1918).

TWENTY

116 **the phenomenal Greek mathematician Demetri Christodoulou:** Here, as in many other places, if a mathematician is referred to by his nickname rather than his given name, it is because the author knows him personally or because everyone calls him by his nickname. In this case both things are true.

120 **(not on a level with Desplechin):** Filbet is referring here to the film director Arnaud Desplechin, whose 2008 film *Un conte de Noël* Villani worked on with Wendelin Werner, helping to write a scene involving mathematical calculations.

TWENTY-ONE

122 *Sleep, little wonders, tomorrow day will dawn:* "Dormez petites merveilles, il fera jour demain"—from the lullaby that the bird sings to his children in the animated cartoon *Le Roi et l'Oiseau*.

TWENTY-TWO

125 **the famous Shanghai ranking:** More formally known as the Academic Ranking of World Universities (ARWU), published each year by Shanghai Jiao Tong University.

127 **"Every mathematician worthy of the name . . .":** From André Weil's memoir *The Apprenticeship of a Mathematician*, translated by Jennifer Gage (Boston: Birkhäuser, 1992), 91.

TWENTY-FIVE

134 **the day of fishes and fools:** An April fool is known in France as a *poisson d'avril.*

134 **watched an episode of *Lady Oscar*:** The animated French version of Ri-yoko Ikeda's popular manga *Berusaiyu no Bara* (The Rose of Versailles).

134 **Gribouille singing "Le Marin et la Rose":** The French singer Marie-France Gaîté, who performed under the name Gribouille, died at a tragically young age in 1968. The video mentioned here can be found at www.youtube.com/watch?v=HmGz-9Ha0MY.

137 **"THE SAILOR AND THE ROSE":** Translation of "Le Marin et la Rose" (lyrics and arrangement by Jean-Marie Huard).

TWENTY-SEVEN

141 **for his spectacular solution of a long-standing problem called the Fundamental Lemma:** Ngô had solved the problem for the case of unitary groups in 2004, and then four years later extended the proof to include Lie algebras. By the end of 2009, even readers of *Time* magazine knew who he was.

142 **as Sigal recommended:** Michael Sigal, a mathematician at the University of Toronto, was a fellow visitor at the IAS for the Spring 2009 term.

TWENTY-EIGHT

157 **The tragic Danielle Messia, the forsaken one:** Villani is thinking here of the hauntingly beautiful song "Pourquoi tu m'as abandonnée?," addressed to an absent father. Messia's music was obsessed with misfortune. She herself died from leukemia while still in her twenties, and has been mostly forgotten since.

157 **spent part of his days devouring *Carmen Cru*:** A series of comic books by Jean-Marc Lelong, published in seven volumes between 1981 and 2001, about a cantankerous old woman living alone in the countryside after the war, cut off from society. A final volume appeared after Lelong's death in 2004.

158 **Jean Ferrat's "Les Poètes":** Based on Louis Aragon's long poem of the same name.

158 **an army captain:** "le vieux con"—the big fool, in Pete Seeger's original lyrics. Allwright recorded a French version of Seeger's 1967 antiwar song "Waist Deep in the Big Muddy" the following year.

164 **"HOLIDAY":** Translation of "Jour de Fête" (lyrics by Catherine Ribeiro/arrangement by Christian Taurines).

165 The photograph of Ribeiro that appears beneath the lyrics hangs on the wall behind Villani's office desk in Paris. The inscription reads: *Je brûlerai jusqu'à extinction des feux* (I will burn until the fires burn out).

TWENTY-NINE

167 **a challenge contemptuously issued by Warren Ambrose:** A distinguished geometer who spent his entire career at MIT, Ambrose deeply resented the verbal abuse heaped on him by an untenured professor almost fifteen years his junior. One day he angrily lashed out at Nash: "If you're so good, why don't you solve the embedding problem for manifolds?" When Nash succeeded ("I did this because of a bet," as he famously said), Ambrose was the first to congratulate him.

167 **the biography by Sylvia Nasar:** Nasar's book, *A Beautiful Mind* (New York: Simon & Schuster, 1998), is the source for the previous note and a number of other details concerning Nash's life and work.

168 *an office in an old factory building just off Washington Square*: The imposing and austere concrete tower that has long been the institute's home at New York University, at 251 Mercer Street, was not built until almost a decade later. The original offices were in a nine-story building at the corner of Waverly Place and Greene Street occupied by hat factories and, on the ground floor, the giant computer being built by the Atomic Energy Commission.

168 *Institute for Mathematics and Mechanics*: Renamed the Courant Institute of Mathematical Sciences after its founder, the German-born mathematician Richard Courant, in 1964.

THIRTY-ONE

176 **mille pompons!:** A favorite expression of Fantômette, the costumed, crimefighting heroine of a series of very popular children's books by Georges Chaulet. It may or may not have anything to do with the fact that Fantômette wears a snug black bonnet with a woolen ball or bobble (a *pompon*, in French) at the end of its long tail.

177 **like the pedestrian in Ray Bradbury's story:** The reference is to Bradbury's 1951 story "The Pedestrian," illuminated in its opening scene by "the faintest glimmers of firefly light."

178 *counties or states or countries that are divided into noncontiguous parts*: In the United States, for example, the Upper Peninsula of Michigan is separated from the lower part of the state by two of the Great Lakes, Michigan and Huron.

179–80 *the research being conducted by an INRIA team*: the French acronym stands for Institut National de Recherche en Informatique et en Automatique (National Institute for Research in Computer Science and Control).

181 **"A word that goes against . . .":** Translation of an excerpt from "La Tour de Babel" (lyrics and arrangement by Guy Béart).

THIRTY-FOUR

190 **Messia's song, Mairowitz's biography of Kafka:** Danielle Messia pays tribute to Prague's revolutionary past in her 1982 song "Avant-guerre." David Zane Mairowitz's *Introducing Kafka* was first published in 1996, with illustrations by R. Crumb.

THIRTY-SIX

202 **From my 2010 Luminy Summer School lecture notes:** The reference is to the summer school held every year at the International Center for Mathematical Meetings on the Luminy campus of the Université d'Aix-Marseille.

THIRTY-EIGHT

208 **Re: Resubmission to *Acta Mathematica*:** With only a few very minor modifications, the letter that follows is the one Villani himself wrote in English.

THIRTY-NINE

211 **one of them the first research monograph I ever read:** Cercignani's *Theory and Application of the Boltzmann Equation* first appeared with Elsevier in 1975.

FORTY

216 **Miyazaki's Nausicaä before the soldiers of the royal house of Pejite:** Nausicaä, princess of the Valley of the Wind, is the title character of the manga series and anime film created by Hayao Miyazaki.

217 **Baudoin's magnificent *Salade Niçoise*:** A graphic novel published by the French artist and illustrator Edmond Baudoin in 1999.

218 **"Those who know do not speak, those who speak do not know":** Schatzman was quoting an aphorism usually attributed to the ancient Chinese philosopher Laozi (Lao Tzu).

FORTY-ONE

221 **the RER:** Short for Réseau Express Régional, a system of five express train lines connecting the center of Paris with outlying suburbs.

FORTY-TWO

226 *Fermat's famous complaint about the margin that was too narrow*: In 1637, Fermat wrote in the margin of his copy of Diophantus's *Arithmetica* that it was too narrow to contain his proof of an adjacent conjecture about positive

integers. The problem resisted solution for more than three centuries, until Andrew Wiles finally succeeded in 1995.

227 *Ages and ages passed*: "Les temps et les temps passèrent"—a line from Catherine Ribeiro's song "L'Oiseau devant la porte."

227 *Thurston was a visionary*: The present tense used in the French edition has been converted here to past since Thurston died in August 2012, just before the book was first published.

227 *chosen by the Clay Mathematics Institute in 2000*: Founded by the American businessman Landon Clay and his wife, Lavinia, the institute sought to concentrate attention on some of the most difficult, and potentially the most rewarding, mathematical challenges of the twenty-first century.

228 *Alexandrov, Burago, Gromov*: Perelman studied with Alexander Danilovich Alexandrov at Leningrad State University and with Yuri Burago at the Leningrad branch of the Steklov Mathematics Institute in the 1980s; in 1991–92 he worked under Mikhail Gromov as a postdoctoral researcher at the Institut des Hautes Études Scientifiques outside Paris.

228 *the so-called soul conjecture*: An outstanding problem of Riemannian geometry first posed by Jeff Cheeger and Detlef Gromoll some twenty years earlier.

FORTY-THREE

231 **number 333 of the one thousand scientists photographed by Maraval:** As part of the exhibition *1000 chercheurs parlent d'avenir*, a video mosaic of faces and words was projected on the walls of the Panthéon in Paris during the week of October 18–24, 2010. A simultaneously published book version is available from www.maraval.org.

231 *Namaste*: The traditional Indian greeting (literally, "I bow to you"), spoken with hands pressed together.

233 **eleven of the fifty-two Fields Medal winners:** The number usually given is ten, but Villani includes Grothendieck, who, though not a French citizen (technically, in 1966, he was a stateless person), received all of his mathematical training in France and worked there for the whole of the time during which he was professionally active. (In 2014, a Fields Medal was awarded to the young Franco-Brazilian mathematician Artur Avila, bringing the total number to twelve.)

233 **a cup of masala chai:** The highly spiced local tea, found in one form or another throughout India and southern Asia.

237 **My contribution to the Korean edition of *Les déchiffreurs*:** A volume of photographs and essays edited by Jean-François Dars, Annick Lesne, and Anne Papillault, originally published in France by Belin in 2008 and subsequently translated into English, Chinese, Japanese, and Korean. The English version given here is new.

237 **TYGER PHENOMENON FOR THE GALERKIN-TRUNCATED BURGERS AND EULER EQUATIONS:** This is the title of the talk given by Frisch at an international conference summarizing the results of a paper subsequently published in *Physical Review* E 84, 016301 (2011) as "Resonance phenomenon for the Galerkin-truncated Burgers and Euler equations." The text that follows is the abstract of the published article.

FORTY-FOUR

240 **My appearance with Franck Dubosc:** A popular comedian in France known for his vulgar humor.

241 **I'd already done an interview with RTL:** Radio Luxembourg, an out-of-country commercial station (originally Radio Télévision Luxembourg) with studios in Paris.

241 **recorded a show for *Des Mots de Minuit*:** A literary program on late-night French television.

242 ***Hasta que el cuerpo aguante!*:** "Until the body gives out!"—a line from a song of the same title by the French songwriter and performer Dominique Ané (better known as Dominique A).

243 **the moment we became aware of his scientific background:** Bernard Accoyer, president of the National Assembly from 2007 to 2012, was trained as a physician.

243 **just before question time got under way:** A lively, sometimes raucous, weekly session during which the government's ministers are submitted to questioning by members of the Assembly; a cousin to Prime Minister's Questions in Great Britain.

244 **the majesty of these extraordinary oversized volumes:** Twenty-three monumental volumes appeared between 1809 and 1828, containing eight hundred thirty-seven engravings in all, many of them larger than any such previous reproductions—and this despite the loss of the expedition's equipment with the sinking of its research vessel near Alexandria in 1798.

A NOTE ABOUT THE AUTHOR

Cédric Villani is the director of the Institut Henri Poincaré in Paris and a professor of mathematics at the Université de Lyon. His work on partial differential equations and various topics in mathematical physics has been honored by a number of awards, including the Fermat Prize and the Henri Poincaré Prize. He received the Fields Medal in 2010 for results concerning Landau damping and the Boltzmann equation.

A NOTE ABOUT THE TRANSLATOR

Malcolm DeBevoise's translations, from the French and Italian, including more than thirty works in every branch of scholarship, have been widely praised. He lives in New Orleans.